스케치와 세계건축여행

Journey of World Architecture & Sketch

merica Mexico Brazil Korea Japan China Taiwan India
hillippines Turkey Australia Spain Italia Greece France United
ingdom Netherlands Sweden Denmark Finland Poland Hungary
zech Austria Egypt Republic of South Africa Kenya Zambia

스케치와 세계건축여행

Journey of World Architecture & Sketch

by Chang Keun, Lee

With the publishment of this book...

Sketch of Le Corbusier seems like a piece of picture with simple crooked shape. While I was looking at his sketches, I discovered that numerous aspects of his experiences of life are hidden in each of the sketches.

In the past, I have sketched as a pathway for my journey of architecture and later I have found that there were only lines without true meaning of architecture. What I have realized truly afterward during my journey of sketches was that "now I really do study about architecture." "If architecture is bowls that holds people, sketch is meditation that holds architecture." This has become my philosophy since then.

Now I dust off my old sketchbooks left in a bookshelf for a very long time and I have combined and published finally all those old sketchbooks containing great many old memories of pleasures and hardships in a single book. It is a new fresh memory and I dream about my new sketch journey that I will begin with professor, Oh, Young-Keun, who is like one of my friends, thinking that this is not the end but it is another new start.

At Hannam-Dong where I can see Mt. Namsan in late fall of 2006

이 책을 내며...

삐뚤삐뚤하게 단순하지만 한편의 그림 같은 르 꼬르뷔제의 스케치, 그 속을 들여다 보고 있노라면 수많은 그의 인생과 경험이 하나하나의 스케치 속에 감춰져 있는 것을 발견하게 된다.

나는 과거에 건축여행을 하면서 한 과정으로 스케치를 해왔는데 그 속에는 선(線)만 존재할 뿐 건축의 진정한 의미를 담아오지 못한다는 것을 후에 알게 되었다. 이후로 내가 진정으로 깨달은 것은 스케치 여행을 하면서 "이제야 건축공부를 하게 되는구나..." 하는 생각이 는다는 것이다. "**건축이 사람을 담는 그릇이라면, 스케치는 건축을 담는 선(禪)**" 이라는 것이 나에게는 철학이 되었다.
이제 책꽂이에 있던 아주 오래된 스케치북의 먼지를 털어내며, 그 사이 수많은 즐거움과 어려움이 담겨있는 모든 스케치를 모아 한 권의 책으로 결실을 보게 되었다. 감회도 새롭지만, 이것이 끝이 아니라 또 다른 시작이라는 생각을 하며 오늘도 친구 같은 오영근 교수와 함께 떠날 스케치 여행을 그려본다.

경술년 늦가을 남산이 보이는 한남동에서

Contents

AMERICA

라스베가스

19세기 말까지는 소규모의 광업과 축산업을 하는 마을이었으나, 1905년에 남(南)캘리포니아와 솔트레이크시티를 잇는 철도가 완성되면서 현대적인 도시로 건설되기 시작하였다. 1911년 3월 16일에 시가 되었다. 1936년에 그 당시로서는 세계 최대의 후버댐이 완성되고, 도박장이 늘어나면서 관광 · 환락지로서 각광을 받게 되어 네바다 주의 최대 재원이 되었다.

Las Vegas

It was a small village of mining and stock farming until the end of 19th century but started to be constructed as a modern city as railroad connecting southern California and Salt Lake City was completed in 1905. I became a city on March 16, 1911. In 1936, the biggest dam at the time, Hoover dam was completed and as gambling places increased and as its popularity of tourism and entertainment attracted people to visit, the city became the biggest income source for the state of Nevada.

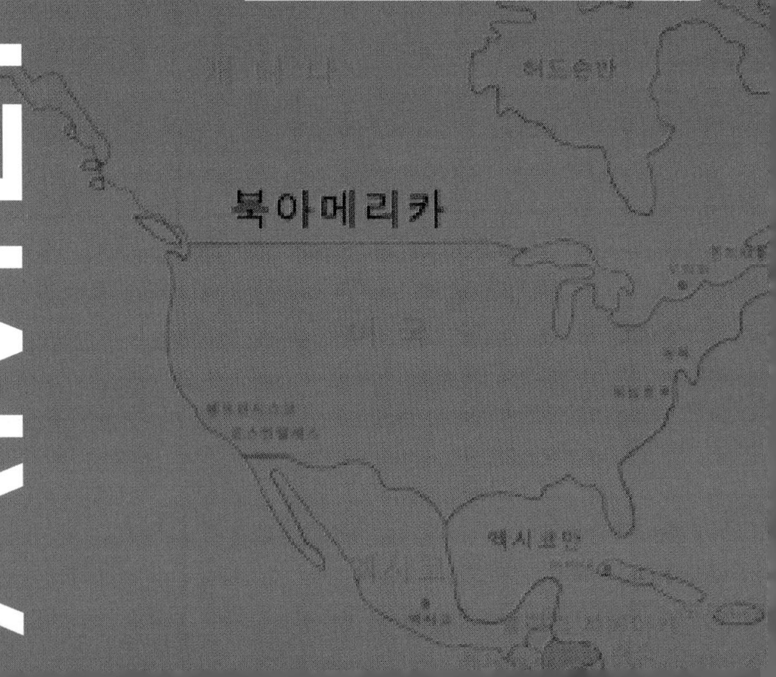

뉴욕

미국 최대의 도시로써 맨해튼, 브룩클린, 퀸즈, 브롱크스, 스테이튼 아일랜드의 5개 독립구로 이루어져 있다. 세계경제의 중심지인 월스트리트, 예술과 문화의 중심지인 브로드웨이, MOMA,구겐하임 등 세계적 박물관, 미술관 등이 곳곳에 있고 세계적 문화도시로도 유명하다. 자유의 여신상, 타임스퀘어 빌딩 등 도시 전체가 관광명소이며 UN본부가 있는 세계적 정치도시라 할 수 있다.

New York

As the largest city in the USA, the city consists of five independent districts of Manhattan, Brooklyn, Queens, Bronx, and Staten Island. The city is famous for the world financial center of Wall Street, arts and cultural center of Broadway, and world famous museum and art museums such as MOMA and Guggenheim and also very famous as world cultural city. The whole city is famous sightseeing place such as Statue of Liberty and Time Square Building and is also the center of world politics with the headquarters of the UN.

로스앤젤레스

1781년 스페인 이주자 44명이 이곳에 오면서 도시가 발전하기 시작했으며, 영화산업과 관광산업, 서비스, 정보산업이 발달된 서부 최대의 도시이다. 면적 약 1,200㎢, 인구 369만 4820명(2000)이다. 뉴욕과 시카고 다음가는, 미국을 대표하는 대도시이며, 주변의 패서디나 · 컬버시티 · 잉글우드 · 산타모니카 · 롱비치 등의 위성도시를 포함한 인구는 700만을 넘어서 뉴욕 지역에 이어 미국 제2의 거대한 대도시권을 형성한다.

Los Angeles

The city started to grow as 44 Spanish settlers arrived here in 1781 and is the biggest city on the west coast of the USA with the development of movie industry, tourism industry, service, and information industry. The total territory is about 1,200㎢ with population of 3 million 694 thousands and 820 people (In 2000). It is a representative city of the USA following New York and Chicago and if population of surrounding areas including satellite cities of Pasadena, Culver City, Englewood, Santa Monica, and Long Beach is included, the population is over 7 million and is the second largest city after New York in the USA.

그리스식 극장

Griffith Observatory in L.A Los Angeles

천문대를 시민과 가까이 할 수 있도록 망원경과 레이져쇼를 통해 밤의 즐거움을 만들어 주고 있으며 그리스식 극장에서는 각종 콘서트를 감상할 수 있다.

It provides the pleasure of nights with telescope and laser shows in order to keep astronomical observatory close to citizen and we can enjoy various concerts in Greece Theatre.

Los Angeles **Schindler House**

프랭크 로이드 라이트와 같이 수학했던 쉰들러가 설계한 이 주택은 미국의 초기 국제주의 양식을 대표하는 건축물 중의 하나이다. 쉰들러는 이 주택에서 특히 화사한 햇살과 따뜻한 열대성의 캘리포니아 기류를 기반으로 한 장소적 접근을 시도하였으나 결과적으로 겨울이나 비가 오는 날에 대한 배려가 부족한 단점으로 남았다. 정확한 디테일과 시공으로 인해 미국에서 건축과 실내디자인을 전공하는 학생들의 실습과 견학의 장으로 즐겨 찾는 공간이 되었다.

The building designed by Schindler who studied with Frank Lloyd Wright is one of representative residences of the early American international architectures. In particular, in this residence, Schindler had tried to make an access to place based on bright sunshines and warm tropical Californian air current but had disadvantage of poor consideration of winter or rainy days as the result. With its precise details and construction works, students majoring in interior designs and architecture visit this house for their on-site practices and study by inspection.

반스달 하우스

Barnsdall House Los Angeles

LA 시내가 내려다 보이는 잔디가 깔린 언덕 위에서 보는 풍경은 아름답기 보다는 마치 소규모 성에 들어와 있는 느낌을 갖는다.

The scenery of Los Angeles viewing from top of the hill of lawns of this house is more than just beautiful but makes people feel like to be inside of a small castle

현대미술박물관

Los Angeles The Museum of Comtemporary Art

1986년 일본 건축가 아라타 이소자키에 의해 디자인된 아주 독특한 박물관이다. 내부는 7개의 갤러리로 구성되어 있으며 이 곳에 몬드리안, 잭스폴락 등의 작품과 마르크스, 로스코, 올덴버그의 팝아트와 추상표현 작품 등을 소장하고 있다. 특히 멀티미디어 기념관에서는 각종 회화, 판화, 사진 등의 작품을 항시 볼 수 있다.

This is a very unique museum designed by a Japanese architect, Arata Isozaki in 1986. The inside consists of 7 galleries and the museum has works of Mondrian, and Jacks Pollack, pop arts and abstract expressions of Marx, Roscoe, and Oldenburg. In particular, we can enjoy various works of pictures, prints, and photography always in multi media memorial hall.

파리호텔

Paris Hotel Las Vegas

프랑스 파리를 연상시키는 에펠탑이 상징적으로 들어오는 내부에까지 에펠탑의 이미지를 끌고 들어오는 독특한 디자인을 나타내고 있다.

The hotel has a very unique design that brings in the image of symbolic Eiffel Tower that reminds Paris into inside of the hotel.

Las Vegas Down Town in Las Vegas

초창기 라스베이가스가 시작된 곳으로 역사적으로 유서깊은 호텔과 추억이 녹아있다. 몰의 천장빛 쇼는 이 곳만의 자랑이다.

This is the place where the initial Las Vegas had begun and there are very old historical hotels and memories. Ceiling light show in the mall is the pride of only this place.

룩소 호텔

Luxor Hotel Las Vegas

이집트 룩소 신전의 이미지를 호텔의 컨셉으로 도입했다. 피라미드 속에서 펼쳐지는 카지노와 식당, 상점 등의 공간 배치가 흥미롭다.

Image of Egyptian Luxor shrine is adopted as the concept of hotel. The space deployment of casino, restaurant, and store inside of Pyramid is interesting.

M.G.M.R 호텔

Las Vegas M.G.M.R Hotel

커다란 코카콜라 병으로 된 속으로 전망 엘리베이터가 보는 이로 하여금 흥미를 갖게 한다.

Observatory tower inside a hugh coca cola bottle makes people attracted and interested.

베네치아 호텔

Venezia Hotel Las Vegas

이태리의 유명 수상도시 베네치아를 벤치마킹했다. 호텔앞에 물을 끌어 들여 베니스에서 볼 수 있는 수상 택시가 관광객의 시선을 끌어 들이고 있다.

It is bench marked as the famous Italian water city of Venezia. Water in the front of hotel attracts sightseeing people like a water taxi that can be seen in Venice.

하라 호텔

Las Vegas Hattahs Hotel

패션쇼 쇼룸

Fashion show Showroom in Lasvegas Las Vegas

라스베이가스에서 볼 수 있는 대규모의 유일한 쇼핑센터이다. 내부에는 각종 디테일 샵이 입주되어 있으며 입구 타원형의 지붕과 대형 디지털 전광판이 흥미롭다.

It is the only large scale shopping center in Las Vegas. Inside the mall, there are many detail shops and oval roof and large digital electric light board at the entrance is very interesting.

스타다스트 호텔

Las Vegas Star Dust Hotel

라스베이가스에서는 비교적 오래된 호텔이다. 외관은 비교적 단순하나 저녁이 되면 보라색의 조명으로 둘러싸인 건물이 무척 화려하다.

It is one of relatively old hotels in Las Vegas. The exterior looks simple but in the evening, it look splendid and beautiful surrounded by violet illumination lights.

교회

Church Las Vegas

신시가지에서 구시가지 쪽으로 걷다보면 조형성이 뛰어난 교회를 볼 수 있다. 많은 돈을 잃고 이곳에서 기도를 하나?

While walking from the new town to old town in Las Vegas, we can see a very excellent church building. Are people praying here after losing lots of money?

플라밍고 호텔

Las Vegas Flamingo Hotel

호텔 입구에 네온사인에서 나오는 불빛이 마치 춤추는 무희를 연상케 한다.
Neon light beaming from the entrance of hotel reminds dancing ladies.

베니스 R.H

Venis R.H in Las Vegas Las Vegas

시저팔래스 호텔

Las Vegas Caesars Place Hotel

라스베이가스에서 가장 고급스럽고 호화로운 호텔로 유명하다. 내부 쇼핑몰 천정의 변화는 마치 옛날 로마의 한 도시를 걷는 듯한 느낌이 들게 한다. 로봇으로 된 로마시대의 스토리 이벤트도 볼만하다.

It is famous as the most luxurious hotel in Las Vegas. The changes of inside ceiling of shopping mall reminds the mood of walking on old streets in Rome. Story event of Roman time in robot is also worthwhile to take a look.

치과병원 MMBB

Dental Clinic MMBB New York

미국 민속박물관

New York American Folk Art Museum

미국의 민속 예술품을 볼 수 있는 작은 규모의 박물관이다. 이 건물의 특징은 단면과 평면의 연속성이 뛰어나다는 것이며, 특히 재료-콘크리트, 나무, 화이버그래스 등이 현대적 감각으로 표현되어 조화를 이룬 디자인으로 잘 나타나고 있다. Bille Tsien은 이 건물을 '작지만 크게 느껴지는 공간' 이라고 극찬했다.

It is a small museum where people can view American folk arts. The characteristics of this building is excellence in the continuity of cross-section and plane surface, in particular, materials of concrete, wood, and fiber glasses are well expressed in modern sense in harmony with design. Bille Tsien has praised this building as 'Small but with the feeling of ample space.'

MEXICO

정식 명칭은 멕시코 합중국이며 수도는 멕시코시티이다. 1521년 스페인에게 정복당했으며 혼란의 시기를 거쳐 1821년 독립이후 공화정으로 자유주의 헌법을 선포하여 오늘에 이르고 있다. 유카탄 반도의 마야 문명으로 이루고 BC 2000년경에 태양과 달의 거대한 피라미드가 지금까지도 그 화려한 역사를 보여 주고 있다.

The official name is the United Mexican States and its capital is Mexico City. It was conquered by Spain in 1521 and became independent in 1821 with democratic constitution through period of confusion and turmoil. It has Yucatan and Mayan civilization and in about BC 2000, they had built huge pyramids of the sun and the moon and they exists until today to show the splendor of history.

MEXICO 멕시코

과달라하라

인구는 164만 7000명(2000)이다. 멕시코 제2의 도시로 해발고도 1,567m에 있으며, 기후가 온화하여 휴양지로서도 유명하다. 1530년경부터 식민이 시작된 곳으로, 그 시대의 경관(景觀)이 남아 있어 '서부의 진주' 라는 별명을 가지고 있다.

Guadalajara

Population is 1 million 647 thousand (In 2000). This is the second largest city in Mexico and is located at altitude above sea level of 1,567m. Climate is mild and also is famous for resort area. This is the place where colonial government began from about 1530 and it has a nickname of 'pearl in the west' as it still keeps those old scenery and landscape.

멕시코시티

인구는 859만 1309명(2000)이다. 정식명칭은 시우다드 데 메히코(Ciudad de Mxico)이다. 에스파냐어(語)로는 메히코라고 한다. 멕시코의 정치 · 경제 · 문화의 중심지이다. 멕시코고원 중앙부의 해발고도 2,240m에 있는 고지(高地)도시로 분지처럼 되어 있는 아름다운 도시이다.

Mexico City

Population is 8 million 591 thousand and 309 people. (In 2000) The official name is Ciudad de Mxico. In Espana, it is called as Mehico. This is the center of Mexican politics, economy, and culture. It is a beautiful city sitting in a basin at the high altitude of above sea level of 2,240m in the central plateau of Mexico.

국제전시장

Expo Guadalajara Mueble International Invierno Guadalajara

과달라하라에 있고 대규모 국제 전시장으로서 언제나 국제적 이벤트와 행사가 시민들에게 제공되고 있다. 이런 전시장의 성격에 맞춰 입구에 짙은 진홍색으로 칠해진 액센트 벽과 이 벽사이로 각종 샵, 음식점 등이 배치되어 있다.

it is a large scale international exhibition building located in Guadalajara and always international events and functions are provided to citizen. In harmony with such nature of exhibition halls, at the entrance, there are accent walls painted in dark red color on both sides and many stores and restaurants are arranged in between.

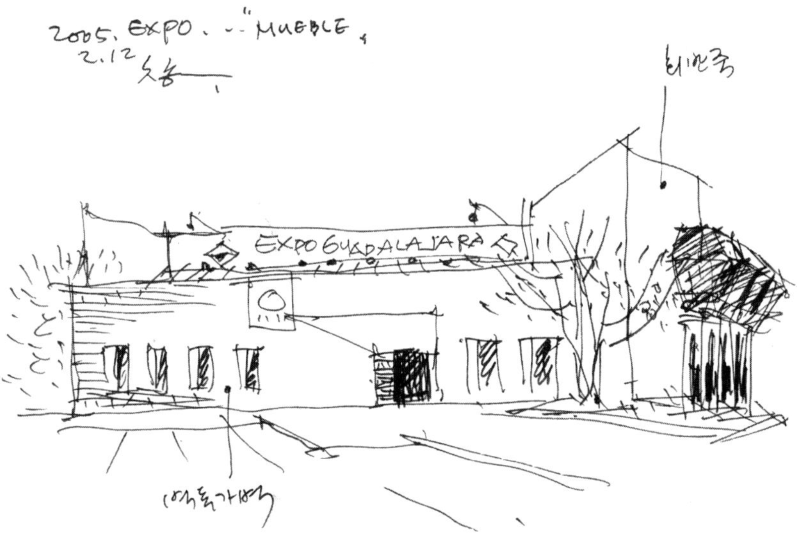
2005. EXPO . . "MUEBLE"
2.12
EXPO GUADALAJARA

호스피시오 카바냐스

Hospicio Cabanas Guadalajara

19세기 초 고아나, 노인 등 어려운 사람들을 돌보기 위한 수용시설이다. 이 건축물의 특징은 수용자들에게 마음의 안식을 주는 교회당의 벽화가 있다.

This was a accommodation facility for taking are of people in difficulties such as orphans and old people in early 19th century. The characteristics of this building is the wall painting that provide spiritual peace for people in accommodation.

CABAÑA
05. 2.
mexico

BACSA Corporation Guadalajara

Mexico City **UNAM University**

라틴 아메리카의 가장 큰 대학으로 대학 내에서 가장 돋보이는 건축물로는 도서관 건물을 들 수 있으며 멕시코의 역사적인 장면을 모자이크로 웅대하게 표현한 Juen corman의 작품이 유명하다.

As the largest university in Latin America, the most eye-catching building in the university is a library building (Bidlioteca Central) and the work of Juen Corman who had expressed a historical scene of Mexico in grandiose mosaic is very famous.

소칼로 광장의 멕시코 대성당

Catedral Metropolitany Mexico City

소칼로 광장에 있고, 200여 년 전에 걸쳐 완공했는데 이 자리는 본래 멕시코인들이 인신공양한 해골들을 모아두던 곳이었다. 본 건물은 1548년 완공됐으나 17세기 들어 남쪽부분이 바로크 양식으로, 북쪽부분이 네오클래식 양식으로 확장돼 웅장하면서도 아기자기한 모습이 하모니를 이루었다. 이 성당의 검은색 피부를 가진 예수상은 유명한 성물이다. 식민정복지에서 원주민을 끌어안으려 노력한 선교의 한 단면이 드려다 보인다.
대성당은 멕시코에서 가장 중요한 역사적 건축물 중 하나이다. 건축은 18세기에 시작되었으며, 이 후 300년 동안 계속되었다. 이렇게 공사를 오래 끌게 된 이유는 건축 양식이 하나로 일관되지 못하고 스페인의 르네상스와 프랑스의 네오클래시즘에 의해 지배되었던 영향 때문이었다. 대성당에 있는 다섯 개의 통로는 다소 야해 보이는 형상과 정교한 조판으로 장식되어 있으며 바닥은 대리석이 깔려있다.

It is located in the Zocalo plaza and is constructed for over 200 years. This was originally the place where skeletons of artificial sacrifice by Mexican were kept. The main building was completed in 1548 but in the 17th century, the southern part was added and expanded in baroque style and the northern part in neoclassic style to give the harmony of grandeur as well as visual charming. Statue of Jesus Christ with black skin in this Cathedral is famous holy artifact. We can feel one aspect of missionary who tried to embrace natives during colonial conquests.
The Cathedral is one of the most important historical buildings in Mexico. The construction began in the 18th century and continued for the next 300 years. The reason for such a long construction period was because the style of architecture was not consistent in one style but was governed by the influence of renaissance of Spain and neoclassicism of France. The five corridors are decorated with more or less gaudy shape and delicate forms and the floor is made of marble.

Catedral Metropolitana
멕시코 대성당

멕시코 대성당
05.2.

후아레스 기념비

Monumento a Juarez Mexico City

멕시코에서 유일한 원주민 출신의 대통령이었던 베니토 후아레스를 기리기 위한 기념비가 멕시코시티 중심부에 세워져 있다. 가난과 출신 성분에 대해 늘 억압 받아온 없는 자들 편에 서서 멕시코의 근대화와 민주화에 기여한 그의 업적은 지금도 멕시코인들에게 가슴깊이 남아 있다.

Monument to remember the only native president in Mexico, Benito Huarez is erected in the mid town of Mexico. His accomplishment of contribution to modernization and democratization of Mexico to be always on the side of people suppressed for their poverty and background of birth is still remembered by Mexican people in their deep heart.

Mexico City Torres de Ciudad Statelite

바라간은 멕시코의 자연과 문화를 건축 어휘로 창조하는데 그것이 그의 인생과 예술, 철학과 함께 공유한다는 점이다.
이 작품은 멀리서 보면 커다란 5개의 덩어리로 보이지만 작품을 하늘을 배경으로 보면 마치 구름 속을 항해하는 배처럼 감성적으로 느껴진다.

Barragan has created the nature and culture of Mexico in the language of architecture and it shares his life, art ,and philosophy.
This work looks like huge five masses if looked from far away but it looks like a vessel sailing into cloud if looked with the background of the sky.

바라간 하우스

Barragan House Mexico City

바라간 건축은 자연을 배경으로 하고 있다. 이 집은 그가 20대에 유럽을 여행한 후 처음 귀국해서 설계한 집이다. 멕시코 전통적 스타일과 내부 수공간으로 된 정원, 독특한 키오스크가 마치 그림처럼 다가온다.

Barragan architecture has the background of the nature. This house was designed after returning first from his travel to Europe when he was in 20s. Mexican traditional style, garden with internal water space, and unique kiosk approach just like shadows.

05.2.13

웨이스 주택

Weiss House Mexico City

건축가 페르난데스의 사무실

Mexico City Office of an architect, Fernandez

멕시코시티 언덕에서 순수기하학의 육면체와 8각의 매스가 독특하게 들어오는 건물을 만날 수 있다. 건축가의 독특한 아이디어와 그의 장난기 어린 계획이 건물의 아이덴티티를 만들고 있다. 절개지 다리를 통해서 들어서면 커다란 철문의 중압감과 대면하여, 수직으로 이루어진 원형 계단을 통해서 사무실에 도달할 수 있다. 아슬아슬한 음식정원과 본인만의 침실도 아주 독특하다.

In a hill of Mexico City, we can see a building with the uniqueness of hexahedron and eight angles in pure geometry. Unique idea and playful plan of the architect makes the identity of the building. When we enter through the bridge of cutting slope, we confront with heavy weight of a huge iron gate and we can arrive in his office through circular stairs made vertically. A dangerous garden and his own bedroom are also very unique.

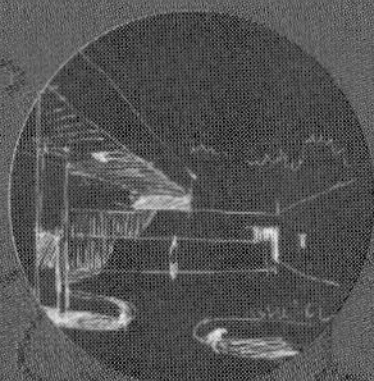
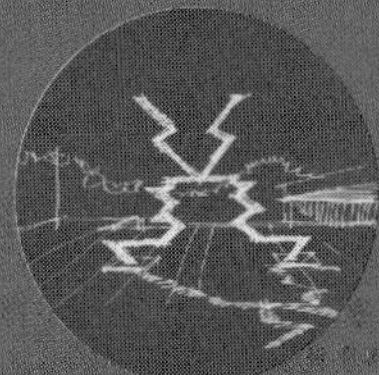

BRAZIL

남아메리카

세계에서 가장 열정적인 나라 브라질은 남미에서 가장 큰 영토와 2억 8천만의 인구(2003)를 갖고 있다. 수도는 브라질리아이며 공용어는 포르투갈어를 사용한다.

The most passionate country in the world, Brazil is the largest country in Latin America and has population of 280 million people. (in 2003). The capital is Brazilia and official language is Portuguese.

BRAZIL 브라질

상파울로

인구는 978만 5640명(2000)이다. 리우데자네이루 남서쪽 약 500km 지점, 해발고도 약 800m의 고원지대에 있으며, 부근의 20여 개 위성도시를 포함하여 인구 900만이 넘는 남아메리카 최대의 도시이다.

리우데자네이루

약칭으로 리우라고도 한다. 인구는 585만 544명(2000)이다. 대도시로 1763~1960년까지 브라질의 수도였으며, 자연미와 인공미의 조화로 세계 3대 미항(美港)의 하나이다. 동쪽은 대서양 연안의 과나바라만(灣)에 면하고, 서쪽은 해발고도 700m가 넘는 가파른 산지가 시의 배경을 이루고 있다. 시가는 코파카바나 · 이빠네마 등 아름다운 해안을 따라 좁고 길게 뻗어 있다.

꾸리찌바

생태도시로 유명한 꾸리찌바는 서울시 청계천 프로젝트를 위한 샘플링이 됐던 도시이다. 잘 정리된 주거단지와 교통 시스템, 친자연적 도시 계획으로 세계의 모범이 되고 있다.

São Paulo
Population is 9 million 785 thousand 640 people. (In 2000) It is located about 500km south west of Rio de Janeiro at high plateau of altitude above sea level of about 800m. Including over 20 satellite cities in the vicinity, it is the largest city in Latin America with the population over 9 million.

Rio de Janeiro
In abbreviation, it is called as Rio. Population is 5 million 850 thousand 544 people. (In 2000). It is a big city and had been the capital of Brazil between 1763 and 1960 and is one of the world' s most beautiful harbor cities with the harmony of the natural and artificial beauty. The east is faced with the Bay of Guanabara in Atlantic ocean and the west is bordered and becomes the background of the city with rugged mountain with altitude above sea level of over 700m. Town streets are stretched in narrow and long shape along beautiful beaches of Copacabana and Ipanema

Curitiba
1 famous as eco city was bench marked by Seoul for the project of Cheonggye river. Well arranged residential area, traffic system, and environment friendly urban planning set a good example of successful city in the world.

꾸리찌바 교회

Curitiba Church Curitiba

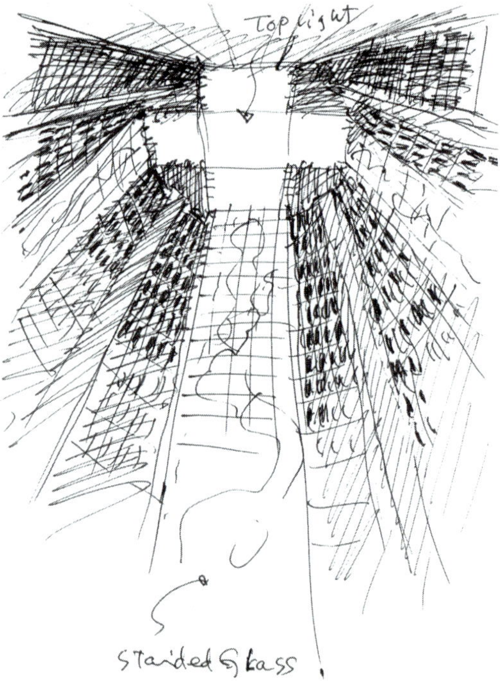

건물 모양의 외관의 매스가 실내에서도 그대로 투영되어 천창의 십자가 모양의 스테인드글라스를 통해 안으로 들어오는 빛이 너무나 황홀하다. 지하에는 중세 전통교회 건축에서처럼 납골당이 배치되어 있다.

External mass shape of the building is reflected inside the building as it is and sunlights coming through stained glass in the shape of cross on the ceiling are so fascinating. In the basement, there is a crypt like traditional churches in the middle age.

Curitiba General Affairs Building

수변으로 넓게 펴져있는 연못위에 환상적으로 건축이 띄워져 배치되어 있다. 내부는 우리가 흔히 생각하는 일반 행정업무를 하는 관공서의 이미지를 탈피하여 유기적인 공간 구성을 하였다.

The building is deployed in the middle of a huge lake fantastically. Inside of the building is not like ordinary office image of general administration of government building but space is arranged systematically.

오스카 니마이어

Museu Oscar Niemeyer Curitiba

브라질의 세계적 건축가인 오스카 니마이어가 설계한 꾸리찌바 명건축물이다. 기하학적 매스와 마치 우주 공간에 떠 있는 듯한 전시장, 하늘과 땅을 연결하는 램프 등은 비중적에서 오는 황홀함에 빠지게 한다.

This is a famous Curitiba building designed by the world famous Brazilian architect, Oscar Niemeyer. Geometric mass and a exhibition space that seems floating in the cosmos, and lamp that connects the sky and the earth make people to be fascinated in non-gravity condition.

Curitiba Children' s regional library (the Lighthouse of Learning)

브라질을 밝히는 원통형의 탑과 2층으로 이루어진 간이 도서관이다. 세계 7대 불가사의 중 하나인 이집트의 알렉산드리아 등대를 본 뜬 '배움의 등대(the Lighthouses of Learning)'를 지난 1995년부터 짓기 시작하여 현재 50곳에 설치되었다. 공동체의 안전을 위한 빛을 발하는 등대라는 개념의 '배움의 등대'에는 5,000여 권의 책을 갖춘 도서관과 독서실이 구비되어 있으며, 많은 시민들이 애용하고 있다. 도서관의 기능뿐만 아니라 저소득층을 위한 교육의 장으로 저녁에는 파출소로서의 기능도 함께 한다.

This is a temporary library consisted of a tower of circular column and two stories that illuminate Brazil. It imitated the Alexandria lighthouse of Egypt that is one of 7 wonders of the world and "the Lighthouses of Learning" were started to be built from 1995 and now there are 50 libraries. As the concept of lights for the safety of local communities, in the 'the lighthouses of learning', there are library with some 5,000 books and reading rooms and are used by many citizens. It serves not only as a library for educating people with low income but also it serves as a police station in the evening.

환경 개방대학원

Environment open university Curitiba

자이메 레르네르가 1992년 건축한 환경교육의 장이다. 친환경적 소재를 원칙으로 통나무 재료로 모든 건축의 구조를 해결하였으며, 램프로 얽혀진 원형의 통로를 따라 강의동과 교육시설이 배치되어 있다.

This is a place for educating architecture environment designed by Jaime Lerner in 1992. All structures of the building are solved by logs in principle of environment friendly building materials and lecture rooms and educational facilities are arranged along corridor of circular column intertwined by lamps.

Curitiba Fangua Park

팡구아 공원은 자연지형을 그대로 살린 채 설계한 공원으로 꾸리찌바 공원사에 있어서 새 지평을 연 장소로 평가받고 있다. 높은 위치에 자리한 입지조건을 이용하여 반원형의 회랑식 2층 건물인 전망대를 만들어 연못과 숲 도시를 바라 볼 수 있게 하였다.
자연적으로 흐르는 물을 이용하여 연못을 이루고 또 이 연못에서 나오는 물은 낙차 큰 지형적 경사를 따라 하부로 유입되어 멋진 폭포를 만들게 된다.

Fangua park is a park designed keeping natural topography as it is and is considered as a place that has opened a new horizon in Curitiba park history. By building a observatory tower of two stories in a half moon corridor style using locational condition of a high hill, people can overview a lake and forest of urban buildings.
A lake is formed using naturally flowing water and water flowing out of the lake makes a beautiful fall by flowing into lower part following the geographical slope of a big difference of water level.

식물원 전시장

Exhibition hall of botanical garden Curitiba

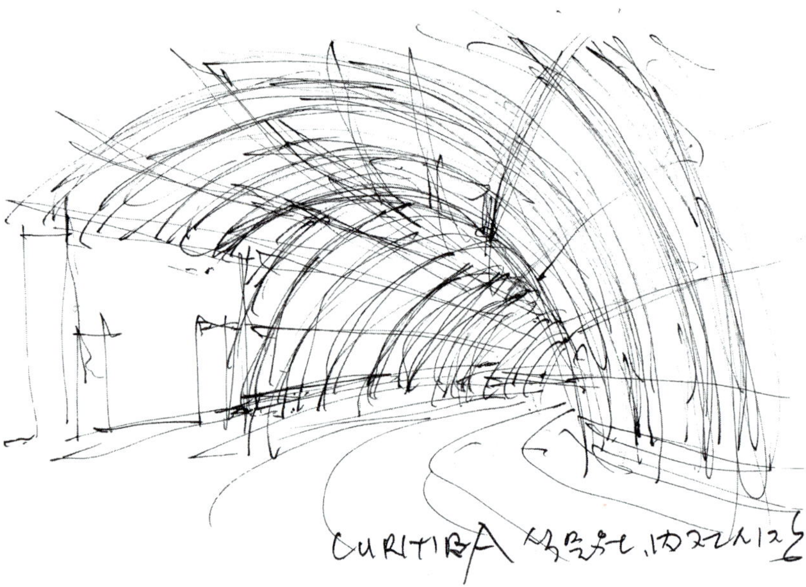

1991년 꾸리찌바에 세워진 프랑스 스타일의 정원이 내부에 있으며 유리로 된 원형 아치로 전시장이 구성되어 있다. 각종 나무로 만들어진 아름다운 형태를 마치 조각품처럼 전시하는 것이 인상적이다. 심지어 화재로 인해 그을린 나무의 앙상한 가지 모양도 훌륭한 조각품처럼 세워져 있다.

A French style garden built in Curitiba in 1991 is located inside and it consists of an arch style exhibition hall made of glass. It is impressive to see various shapes made of wood exhibited just like sculptures. Even a tree with bare branches smoked in a fire is erected as a wonderful sculpture work.

Rio de Janeiro Museum of Contemporary Arts

1993년 리우데자네이루의 아름다운 해변에 세워진 미술관이다. 총 3층으로 구성되어 있으며 1/2로 된 지하 공간에는 전시공간, 컬렉션 공간이 있으며, 3층에는 오피스와 레스토랑 그리고 70석의 강당이 있다.

This is a museum built at a beautiful beach of Rio de Janeiro in 1993. It consists of the total 3 stories and in a half basement space, there are exhibition and collection spaces and in the third floor, there are offices, restaurants, and an auditorium with 70 seats capacity.

선착장

Dock Rio de Janeiro

리우데자네이루의 항구 입구에 고전적 스타일의 작은 건축물이 눈에 띈다. 과거 포르투갈의 지배 때 이곳을 이용하는 왕을 위한 선착장이다.

At the entrance of Rio de Janeiro harbor, a small classical building is eye-catching. This is a dock for kings who used this place during Portuguese colonial period.

코파카바나 해변

Rio de Janeiro Copacabana

리우데자네이루 내에서 가장 유명한 해변이 코파카바나이다. 너무나 정열적이다 못해 이곳에서의 독특한 비키니를 입은 여인들의 수영복 패션은 전 세계에서 몰려오는 관광객에게 흥미와 정취를 느끼게 한다.

The most famous beach of Rio de Janeiro is Copacabana. It is too passionate and swimming fashion of women wearing unique bikinis is world famous to attract and instigate interest and sentiment of tourists from all over the world.

Museu de Arte São Paulo São Paulo

1947년 Assis Chateaubriand가 건립하였으며, 현존 건물은 1968년 Lina Bo Bard가 80m 길이의 2층 건물을 지상은 비워둔 채 4개의 기둥만 붙이고 설계한 것으로 유명하다.

It was built by Assis Chateaubriand in 1947 and the current building is famous as Lina Bo Bard had designed in 1968 with only 4 pillars in the two stories building of 80m high with empty ground.

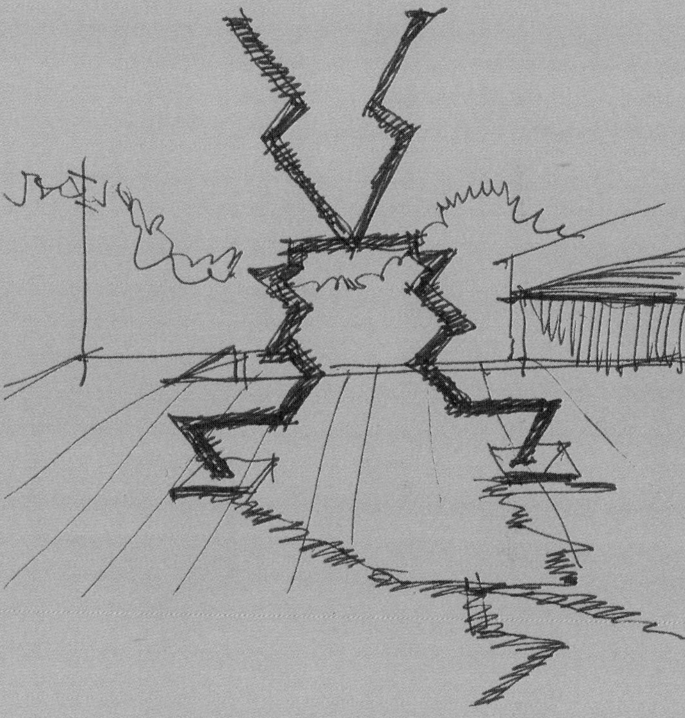

JAPAN

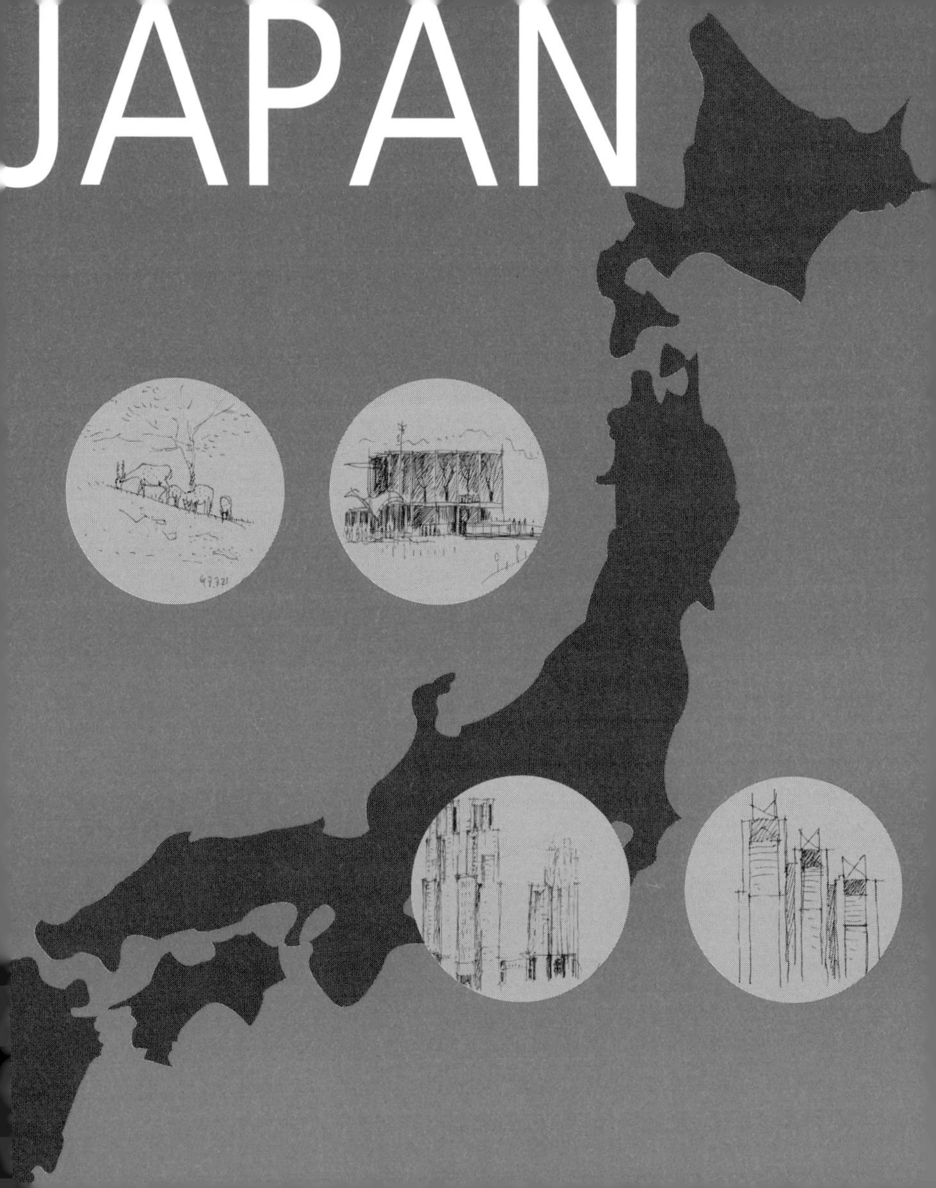

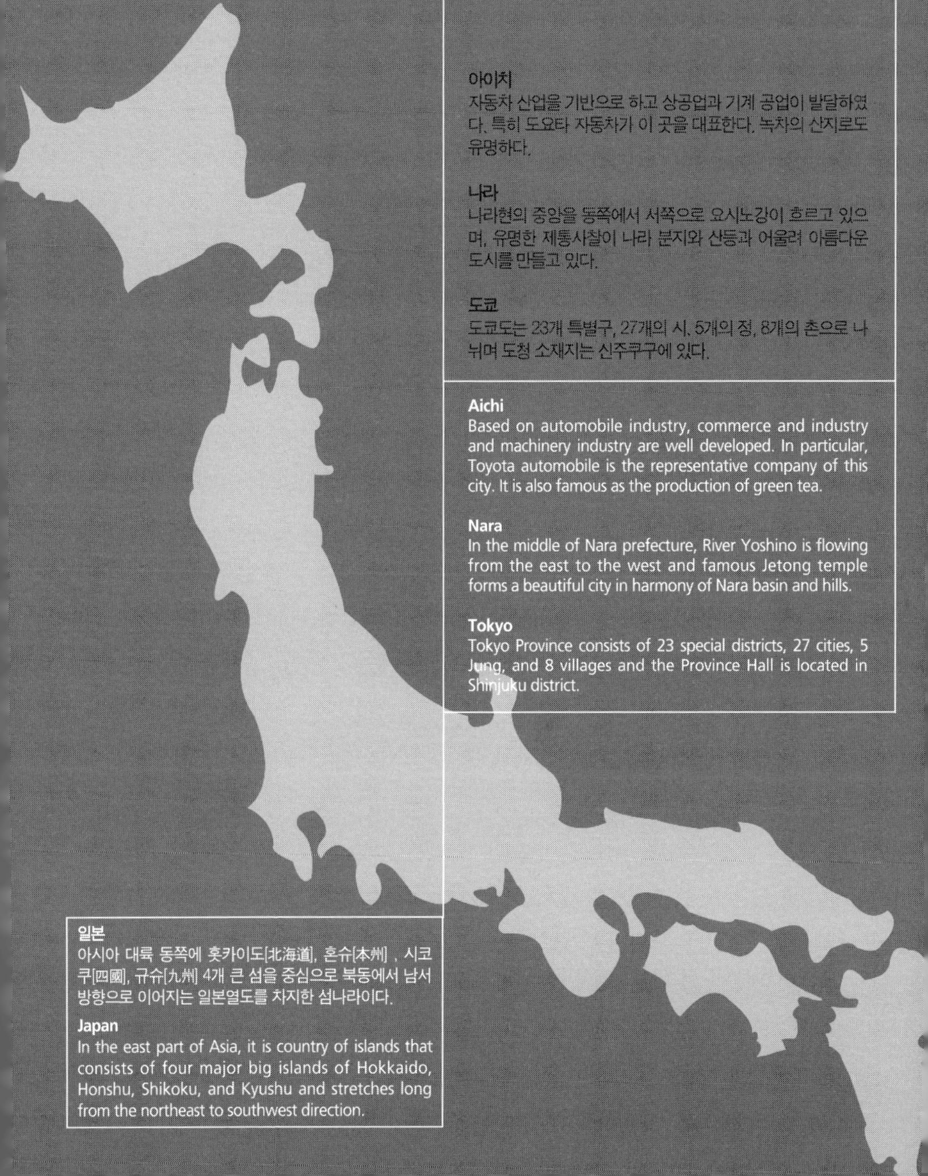

아이치
자동차 산업을 기반으로 하고 상공업과 기계 공업이 발달하였다. 특히 도요타 자동차가 이 곳을 대표한다. 녹차의 산지로도 유명하다.

나라
나라현의 중앙을 동쪽에서 서쪽으로 요시노강이 흐르고 있으며, 유명한 제통사찰이 나라 분지와 산등과 어울려 아름다운 도시를 만들고 있다.

도쿄
도쿄도는 23개 특별구, 27개의 시, 5개의 정, 8개의 촌으로 나뉘며 도청 소재지는 신주쿠구에 있다.

Aichi
Based on automobile industry, commerce and industry and machinery industry are well developed. In particular, Toyota automobile is the representative company of this city. It is also famous as the production of green tea.

Nara
In the middle of Nara prefecture, River Yoshino is flowing from the east to the west and famous Jetong temple forms a beautiful city in harmony of Nara basin and hills.

Tokyo
Tokyo Province consists of 23 special districts, 27 cities, 5 Jung, and 8 villages and the Province Hall is located in Shinjuku district.

일본
아시아 대륙 동쪽에 홋카이도[北海道], 혼슈[本州] , 시코쿠[四國], 규슈[九州] 4개 큰 섬을 중심으로 북동에서 남서 방향으로 이어지는 일본열도를 차지한 섬나라이다.

Japan
In the east part of Asia, it is country of islands that consists of four major big islands of Hokkaido, Honshu, Shikoku, and Kyushu and stretches long from the northeast to southwest direction.

일본 국제 박람회

The 2005 Japanese World Expo Aichi

「EXPO 2005 AICHI, JAPAN」는 '자연의 예지' 를 테마로, 21세기의 자연과 인간의 관계를 제시주제로 일본의 아이치에서 열렸다. 인류가 직면하고 있는 생활과 공존의 여러 문제들을 최첨단의 기술과 시스템으로 이를 해결할 수 있는 가능성을 제시하였으며, 사람들에게 이를 직접 체험시킴으로써 함께하는 박람회를 추구하였다.

「EXPO 2005 AICHI, JAPAN」was held in Aichi, Japan under the subject of the relationship between the nature and people in the 21st century with the theme of 'wisdom of the nature.' Possibility of solving various problems of life and coexistence confronting human being was presented through advanced technology and system and people could experience this in person as an Expo with participation of people.

Canada
Expo 2005
AICHI
JAPAN

일본 국제 박람회

The 2005 Japanese World Expo Aichi

나라 공원

Nara **Nara Park**

나라 시 동부의 나라 공원은 동서 약 4km, 남북 약 2km의 공원으로, 원내에는 도다이사 절, 고후쿠지 절, 가스카타이샤 등, 일본을 대표하는 절과 신사가 있다. 아름다운 경관사이로 사슴이 사람과 함께 뛰어 노는 모습이 무척 자연스러워 보인다.

Nara park located in the east of Nara City is 4km long in the east and west and about 2km long in the north and south. Within the park, there are representative temples and shrines of Japan such as Todaida temple, Kohukuji temple, and Gaskatai. It looks so natural that deers and people are playing among beautiful sceneries.

신주쿠 파크 다워

Shinjuku Park Tower Tokyo

지상 235m, 52층 건물로 지하 1층에는 '파크타워 에비뉴' 라는 레스토랑 거리가 형성되어 있다. 3~8층까지는 인테리어 등의 주거 공간을 테마로 한 리빙 디자인 센터인 '오존' 이 자리를 잡고 있으며 39~52층까지는 '파크 하얏트 도쿄' 호텔이다.

일본 초고층 빌딩 중에서도 최대 기준단면적을 자랑하는 파크타워는 그 창조력이 최대한으로 발휘된 건물 중의 하나이다. 기본적인 모티브로서 인간 중심의 여유있는 공간을 구현하고 있다.

시각적인 편안함을 주는 무기둥(無柱) 설계, 최신 OA기기의 마루 배선시스템 등 쾌적성과 편리성을 강조한 건물이다.

It is 235m high and consists of 52 stories. In the first basement, there is 'Park Tower Avenue' restaurant street. In 3~8 floors, there is 'Ozone' that is a living design center with the theme of residence space like interior and it is 'Park Hyatt Tokyo' Hotel between 39~52 floors.

Park tower has the highest unit area among super high-rise building in Japan and is one of buildings with the maximum realization of creativity. As a basic motif, it pursues human-centered surplus space.

This is a building with no column design for visual relax and the emphasis of pleasure and convenience through floor wiring system of advanced OA devices.

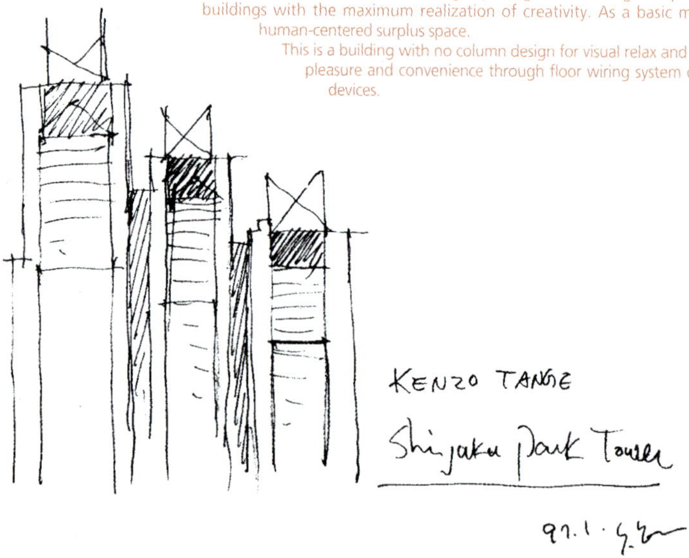

-TOKYO 2004.12.

휴멕스 파빌리온 빌딩

Humax Pavilion Building Tokyo

시부야 번화가에서 약간 작은 도로변 언덕에 다다르면 마치 우주선 같은 이 건물을 만날 수 있다. 건축가 와카바야시 히로유카의 작품으로 하이테크한 건물 매스의 이미지가 사람들의 시선을 사로잡는다.

You can meet this building looking like a spaceship if you reach to a small hill side walking from busy business quarter in Sibuya. This is the art work of an architect, Wakabayashi Hiroyuka and the image hightech building mass attracts attention of people.

동경시청사

Tokyo Tokyo City Hall Metropolitan

일본 근대 건축의 우상으로 세계적으로 유명한 겐조 단게가 설계한 시청 건물이다. 일본의 전통과 최첨단의 테크닉이 잘 조화되어 동경 시내의 랜드 마크가 되고 있다. 시청 건물 맨 꼭대기에 위치한 전망대에 오르면 모든 동경 시내가 한눈에 들어온다.

This is a city hall designed by the world famous Kenjo Dange as an idol of modern Japanese architecture. It has become a landmark of Tokyo City in a great match of Japanese tradition and the most advanced technology. In the observatory tower on top of the city hall building, we can overlook the whole sceneries of Tokyo City.

CHINA

중국(中國)이라는 국명은 '중심문화' 라는 의미로 중화(中華)와 같은 뜻을 갖는다. 워낙 방대한 면적(957만 2900㎢)과 12억을 넘는 인구에서 보듯이 중국이 아시아에서 갖는 지위와 역사성은 막강하다 하겠다. 최근 개방의 흐름을 타고 경제적으로도 아시아를 넘어 세계적으로 중요한 위치를 차지하고 있으며, 상해, 북경, 심양 등에 세계적 명성의 건축가들과 디자이너들의 작품이 곳곳에 세워지고 있다.

National name of China means the 'central culture' and has the same meaning of Junghwa. As seen in such vast area (9 million 570 thousand and 2900㎢) and population over 1.2 billion, the position and history of China in Asia has significant and powerful meaning. With the recent flow of opening, it becomes as an important part of the world in economy going beyond Asian market and there numerous works of architectures built by the world famous architects and designers in Shanghai, Beijing, and Shenyang.

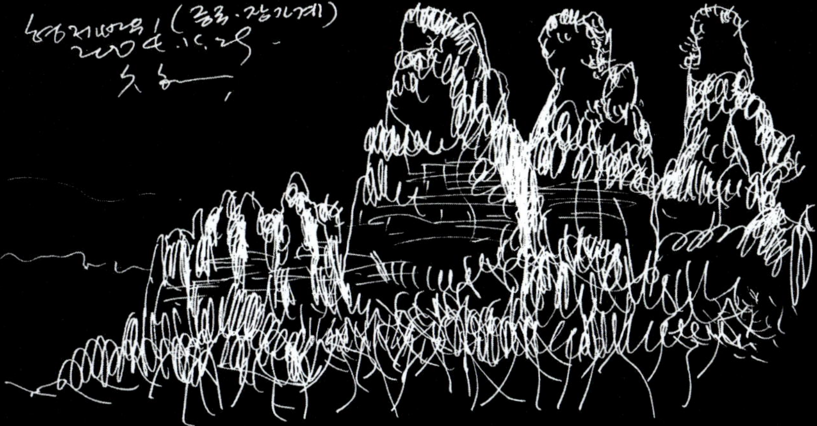

심양

옛 이름은 봉천(奉天), 만주어명은 무크덴(Mukden)이다. 둥베이[東北]지방 최대의 도시로 이 지방의 정치·경제·문화·교통의 중심지이다. 둥베이 남동부의 노년기산지 말단부가 평야와 접하는 랴오허강[遼河] 유역에 있다.

Shenyang

The old name is Bongcheon and its Manchurian name is Mukden. This is the largest city in the northeast region and is the center of politics, economy, culture, and transportation. The end part of old mountain region in the northeast is faced with the plain of River Raohyeo.

상하이

역사적으로도 중국의 문화, 경제의 중심 도시로써 중요한 위치를 차지하고 있다.

Shanghai

As a central city of culture and economy in China historically, the city has an important position in China.

서안

서안은 주(周)나라 무왕(武王)이 세운 호경(鎬京)에서 비롯되며, 그 뒤 한(漢)나라에서 당(唐)나라에 이르기까지 약 1,000여 년 동안 번영한 역사적 도시이다. 진시황제가 BC 247년 즉위하면서 만든 자신의 묘가 세계에 알려지면서 더 유명해졌다.

Cyanogen

Cyanogen had began from Hogyeong city built by King, Mu of Ju Dynasty and had prospered for over 1,000 years in Chinese history from Han dynasty to Dang dynasty. Jinsi Emperor was inaugurated in BC 247 and built his tomb. As his tomb was known to the world, it became more famous.

대안탑

Goose Pagoda Cyanogen

높이 64m, 7층의 이 탑은 바닥이 정방형이며, 사람이 탑 안에 들어갈 수 있는 서안의 대표적 상징물이다. 이 탑은 삼장법사 현장이 인도에서 불도로 통해 얻은 중국어로 된 불경을 보존하기 위해 건립되었다.

This is a representative symbol of Cyanogen with the height of 64m in 7 stories. The foundation of the pagoda is regular square and people can go inside the pagoda. The pagoda was built to preserve Buddhist Scriptures in Chinese by Hyeonjang, a Buddhist sage of Samjang. He had obtained the scriptures in India through his meditation training of Buddhism.

웨스트 게이트

Cyanogen **West Gate**

서안은 과거 중국의 수도로써 그 위용이 대단했으나, 지금은 북경, 상하이 등 동쪽에 비해 낙후되어 있다. 따라서 서안 곳곳에는 오래된 건축물과 유적이 곳곳에 산적해 있어 관광객에게 마치 옛날로 돌아간 것 같은 착각에 빠지게 한다.

Cyanogen was powerful as the old capital of China but now it falls behind the east such as Beijing and Shanghai. Therefore, there are various old ruins and buildings in Xi' an and makes tourists feel that they are returned to the old time.

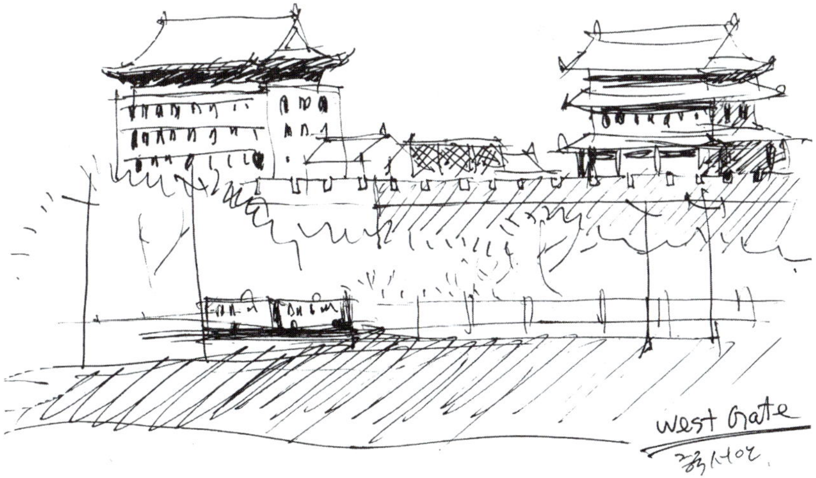

진시왕 병마용

Qin Shi Huang, Terra-Cotta Army Cyanogen

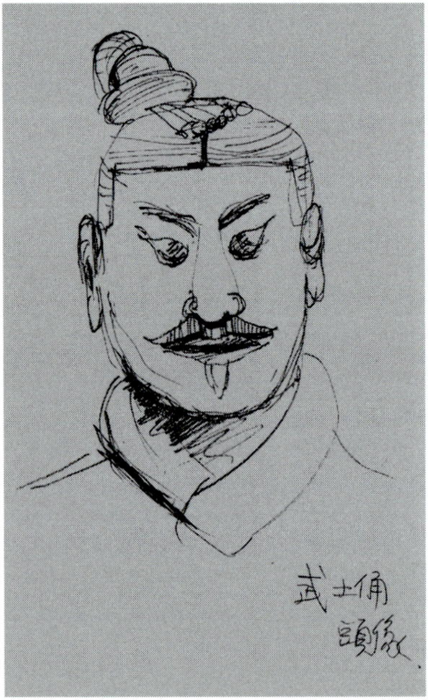

1947년 처음 발굴이 시작된 후 지금까지 3호갱까지 연차적으로 발굴되었다. 1호갱은 길이가 210m, 너비 60m깊이 4.5~6.5m의 총 12,000m²이며 약 6000여개의 병마용이 있다. 2호갱은 총6000m² 정도이며 전차를 끄는 마용이 356개 기병용이 116개, 보병용이 562개 발견되었다. 3호갱은 총 520m²로써 1, 2호갱과는 달리 지휘부로 추정되는데, 갑옷입은 보병용64개, 마용4개와 장군용으로 보이는 채색된 전차 1개가 발굴되었다. 이런 병마용들은 거의 1.5m~2m정도에 이르는 실물크기의 모습이며 각 무사의 얼굴의 표정, 심지어 손금까지도 전부 상이하게 표현되어 얼마나 정성을 들여 만들었느냐를 짐작케 한다. 진시황 병마용은 그 당시의 군사 편제 및 의복, 무기 연구에도 큰 자료가 되고 있다.

It was first excavated in 1947 and until now total three tunnels were excavated gradually. The length of the first tunnel is 210m and the width of 60m with the depth of 4.5~6.5m in total area of 12,000m² and there were about 6000 soldiers and horses terra cotta. The second tunnel is about the size of 6000m² and 356 chariot horses, 116 cavalries, and 562 foot soldiers were discovered. The third tunnel has the total area of 520m² and is presumed as headquarters unlike the number 1 and 2 tunnels and 64 armored foot soldiers, 4 cavalries, and colored chariot that seems like for general were discovered. These soldiers and horses are in the height of actual size of 1.5m~2m and facial expression and even palm lines are expressed differently in details per soldier and makes us presume how carefully all these terra cottas were produced. Qin Shi Huang' s terra-cotta army become very important information for studying the army organization, clothes, and weapons of that time.

장가계(張家界)

Janggagye (Family Jang World) Cyanogen

맑은 물과 깊은 협곡, 기이한 산봉우리 등 천혜의 자연이 그림처럼 펼쳐진다. 장가계는 중국 호남성 서북부에 위치하며, 아열대 기후와 생물 생태 환경으로 관광객을 끌어 들이고 있다.

The natural bounty of clean water, deep valley, and grotesque mountain peaks are stretched like a picture. Janggagyw is located in the northwest of Honam Province of China and attracts tourists with subtropical climate and creatures eco environment.

어필봉(御筆峰)

Eopilbong (Royal writing peak) Cyanogen

황제가 쓰던 붓을 던진 형태라 해서 '어필봉' 이란 이름이 붙여졌다. 서있는 서로 다른 3개의 봉우리 위에 푸른 소나무가 자라서 보는 사람으로 하여금 붓을 꽂아 논 형태를 연상하게 한다.

The name of 'Eopil peak' was given as the shape is like that a brush used by Emperor was thrown away. Green pine trees are grown on three differently erected peaks facing one another and makes people remind seeing the shape of brushes put in a case.

세 자매 바위

Cyanogen Three sisters rock

장가계내에 모노레일의 종착역 쪽에 있는 세 자매 바위는 첫째와 둘째 바위는 아이를 업고 있는 모습이며 셋째 바위는 임신을 한 듯 배가 불러있는 모습이다.

Three sisters rock located at the destination of monorail within Janggagye are in the shapes of sisters holding babies on their back for the first and second rock sisters and the shape of pregnancy for the third rock sister.

화청지(華淸池)

Hwacheongji (Bright and clear lake) Cyanogen

화청지는 양귀비와 현종이 겨울을 보내던 별장으로 중국에서 현존하는 최대의 당대 왕실 원림이다. 양귀비가 실제 목욕을 했다는 목욕탕이 있으며, 주변의 수려한 풍경과 약효가 있다는 온천수로 인해 중국에서뿐 아니라 세계적으로 유명해졌다.

Hwacheongji is famous as Queen Yangguibi and King, Hyunjong spent in winter and is the largest royal forest existing now in China. There is a bathtub where Queen Yangguibi took bath and it has become famous not only in China but also in the world because of beautiful natural sceneries and medically effective hot spring in the vicinity.

시안. 화청지 에서
04.11.28

부용진(芙容鎭)

Buyongjin Shenyang

영화 "부용진"의 배경이 되었던 마을로써 그 거리와 집의 보습을 통해 영화 속의 이미지를 연상케 한다.

This is the village of the background of the movie "Buyongjin" and all streets and houses were remodeled to remind the images in the movie.

상하이 푸동지역

Shanghai Building Group of Pudong New Area

'상하이' 하면 옛날 영국과 청나라의 아편 전쟁의 무대로써 역사적 배경을 갖고 있지만, 지금은 국제도시로서 뿐만 아니라 중국 최대 무역항으로써 세계경제의 한 축을 담당하고 있다.

'Shanghai' reminds people as the stage of historical opium war between Great Britain and Chinese Cheong Dynasty but now it is an international metropolitan city and plays an important role in the world economy as the largest port of trade in China.

상해 위스틴 호텔

Westin Shanghai Hotel Shanghai

호텔내 높이 26.5m되는 로비 아트리움은 세련되고 웅장하며 그중에서 4개 층을 연결하는 유리로 된 엘리베이터는 디자인이 독특하다.

Lobby atrium with the height of 26.5m within the hotel is elegant and grandiose and among them, elevators made of glasses that connect 4 floors are very unique in its design.

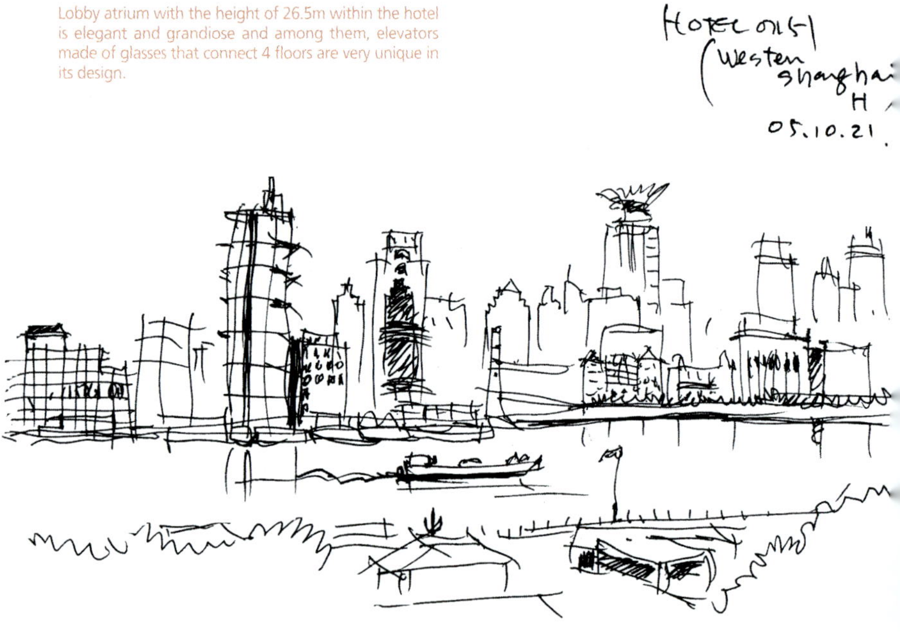

Shanghai Bank of China Tower

발전하는 중국 경제의 상징으로 중국 은행은 상하이 중심부에 대표적 은행 건물로 우뚝 서있다.

This is a symbol of growing economy of China and Bank of China Tower is erected in the middle of Shanghai as a representative bank building.

투모로우 스퀘어

Tomorrow Square Shanghai

7~35층은 아파트로, 36층은 기계실, 37층 이상은 호텔이며, 타워와 포디움의 2개의 영역으로 구분되어 있다. 타워의 스퀘어 플랜은 37층에서 47도 각도로 회전되어 호텔의 시작점이 된다.

7~35 floors are apartments, 36th floor is machinery room, and floors above 37 are hotel and it is divided into two zones of tower and podium. Square plan of tower is twisted in the degree of 47 angle on the 37th floor to begin as a hotel.

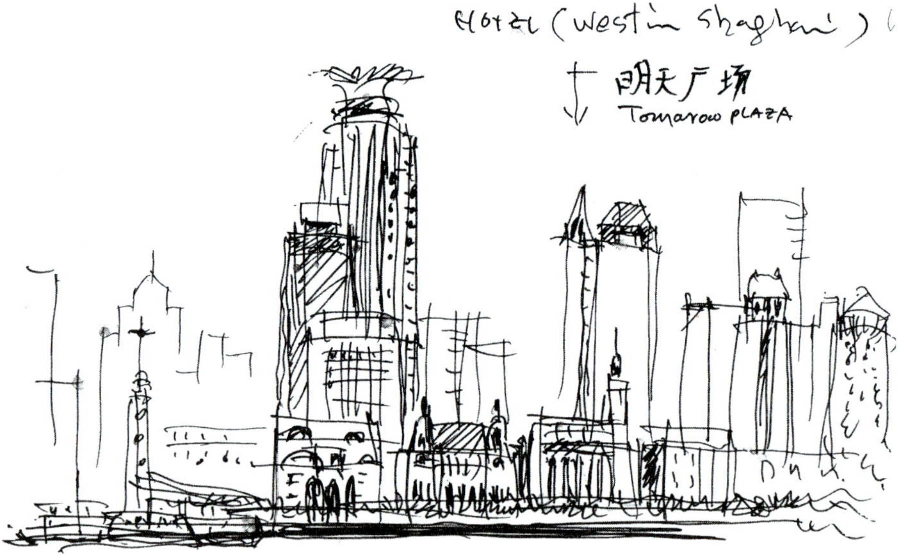

플라자 66

Shanghai Plaza 66

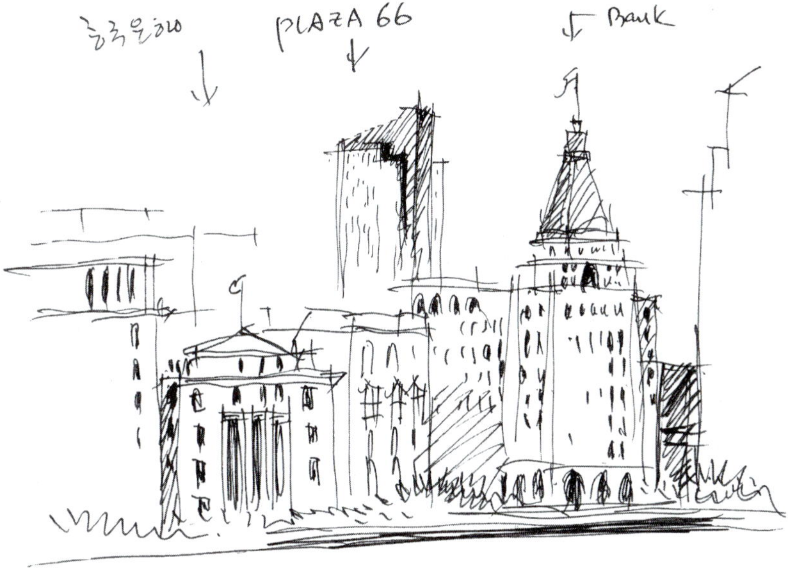

건물은 66층과 46층 2개의 오피스 타워로 구성되어 있으며 그 사이 5층의 포디움이 자리잡고 있다. 46층 저층 타워가 있으며 포디움은 대부분 리테일 공간으로 구성, 세계 유명 브랜드들이 입점해 있다.

The building consists of two office tower of each 66 stories and 46 stories and 5 stories of podium are located in between the lower 46 stories tower and in the podium, there are many retail stores of the world famous brands.

진마오 타워

Jinmao Tower Shanghai

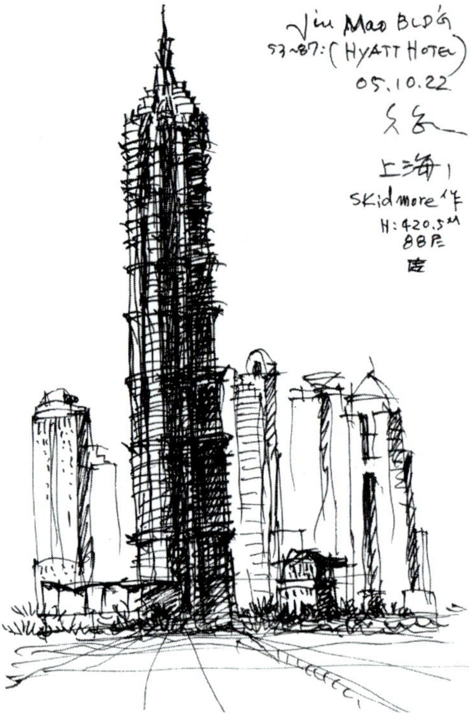

미국의 세계적 건축 설계회사 SOM이 설계한 이 건물은 1993년 현상설계에 들어가 1998년 완공된 88층의 초고층 복합빌딩으로, 상하이 푸둥[浦東]지구에 있다.

전체건물의 공간을 보면 50층까지는 업무시설이며, 51층부터 88층까지는 호텔이다. 타워 이름인 '진마오'는 중국어로 '많은 돈'을 뜻한다. 중국의 시장개방과 자본주의의 영향을 잘 보여주는 예라고 할 수 있다. 또 건물의 설계에서부터 시공에 이르기까지 모든 것이 숫자 8과 관련되어 있는데, 층수 88층, 높이와 너비의 비 8대 1, 전단벽 코어(core) 8면체, 외곽 복합기둥 8개, 완공일 1998년 8월 8일 등이 그것이다. 이는 중국인들이 숫자 가운데 8을 선호하기 때문이다.

The building is designed by the world famous architecture company of America, SOM and first went into competition drawings in 1993 and was completed in 1998 in 88 stories high-rise complex building and is located in Pudong district in Shanghai.

Of the total space of the building, up to 50th floor, it is business facility and from 51st to 88th floor, it is a hotel. The name of tower, 'Jinmao' means lots of money in Chinese. This is a good example of the market openness and the influence of capitalism in China. In addition, from the design to the construction works of the building, everything is related to the number 8. The number of floors of 88, the ratio of the width and the height of 8:1, 8 sides in the front wall core, 8 outside complex columns, and the completion date on August 8, 1998. This is because chinese like the number 8.

동방명주타워

Shanghai Oriental Pearl Tower

상하이에서 가장 상징적인 동방명주(東方明珠)는 황포강변의 포동 개발지구에 위치하고 있다. 원래 방송 관제탑인 동방명주는 높이가 무려 468m나 된다. 이 수신탑은 크고 작은 11개의 구(球)로 이루어져 있는데 구는 진주를 황포강은 옥쟁반을 상징하며, 전체적으로 크고 작은 진주가 옥쟁반에 떨어지는 이미지를 표현하고 있다.

As the most symbolic in Shanghai, Oriental Pearl Tower is located in Pudong development district alongside of River Hwangpu. Originally it is a broadcasting tower with the height of 468m. This receiving tower consists of 11 small and big globes and globe symbolizes pearl and River Hwangpu symbolizes a jade plate and as the whole, it shows the image that small and large pearls fall onto a jade plate.

오리엔탈 아트센터

Oriental Art Center Shanghai

세계적인 프랑스 건축가 폴 엥드류가 설계했으며, 2005년에 완공되었다. 외관은 5개의 꽃잎 모양으로 만개했을 때는 호접란을 상징하며, 5개의 꽃잎은 정문, 오리엔탈 퍼포먼스 홀, 오리엔탈 콘서트 홀, 전시실, 오리엔탈 오페라 홀 등을 상징한다. 외벽은 푸른 강화 유리로 처리했으며, 잘 짜인 철재 프레임이 밖으로 돌출돼 있어 현대적인 감각을 드러낸다. 정문에서 공연장으로 향하는 계단 양편과 통로 벽에는 어른 손바닥만한 돌 수 천 개로 장식했다.
오리엔탈 콘서트홀은 1,353석으로 자석 배치가 여러 개의 언덕 모양을 하고 있어 마치 무대가 언덕 사이에 숨은 계곡을 연상시킨다. 오리엔탈 퍼포먼스 홀은 333석으로 자석이 계단형이며, 반원으로 퍼져 나가는 모습이다. 로마시대의 극장을 본뜬 것으로 독창회, 챔버 콘서트, 미니 드라마가 열린다. 오리엔탈 오페라 홀은 붉은 색 바닥이 인상적이다. 이탈리아 공연장을 보듯 벽면에도 자리가 있으며, 2층으로 나뉘어 있다. 자석은 1,015석이며, 무대 면적은 120㎡이다. 100명 이상의 뮤지션들이 연주를 할 수 있으며, 컴퓨터로 통제되는 다양한 첨단 무대장치가 설치돼 있다. 주로 오페라와 발레, 뮤지컬이 공연된다.

It is designed by the world famous french architect, Paul Andrew and was completed in 2005. Flower petals in the exterior symbolize orchid in full blossom and each symbolizes main entrance gate, oriental performance hall, oriental concert hall, exhibition, and oriental opera hall respectively. External wall is treated with blue reinforced glasses and well woven steel frames are extruded to provide modern feeling. Both sides of stairs from the main entrance gate to the performance hall and the walls of corridor are decorated with several thousands of pebbles in the size of adults' palms.

Oriental concert hall has 1,353 seats in tiers like several hills and the stage reminds the valley hidden between hills. Oriental performance hall has 333 seats in stair style and extends like a half moon. It imitated theaters of Roman period and solo concert, chamber concert, and mini drama are held. Oriental opera hall has unique red color floor. Like an Italian performance hall, it has seats on the side of wall and is divided into 2 stories. Seat capacity is 1,015 seats and the area of stage is 120㎡. Over 100 people of performers can play and various advanced stage devices controlled by computer are installed. Mainly opera, ballet, and musical are performed.

2004
Oriental ARTCenter 上海
05.10.22.
— 연꽃모양 —
Paul Andreu
— ECAD

과학기술박물관

Science & Technology Museum Shanghai

미국의 RTKL 건축 사무소에서 설계한 전기통신박물관과 통신설비실, 사무실 등의 기능을 통합한 빌딩으로 중국 통신 산업의 상징적 건물이다. 높이는 288m이며 지상 41층, 지하 4층으로 이루어져 있으며 통신 설비실이 위치한 저층부의 경우 평면은 직교하는 사각박스이며 입면은 가늘고 긴 창문으로 구성, 유리로 마감한 고층의 사무실 존과 대비를 이루도록 하였다. 박물관의 전시공간으로 사용되는 아트리움은 35m 높이의 슈퍼트러스에 매달린 구체와 유리박스를 포함한다.

This is a symbolic building of Chinese telecommunication industry as a comprehensive building of electrical communication museum, communication facility room, and offices designed by an American RTKL architecture company. The height is 288m and consists of 41 stories above the ground and 4 stories underground and it provides clear comparison of lower floors with communication facility are in crossing rectangular box and narrow and long windows with high-rise offices with glasses. Atrium used as exhibition space of museum includes globe and glass box hanging in 35m height on super truss.

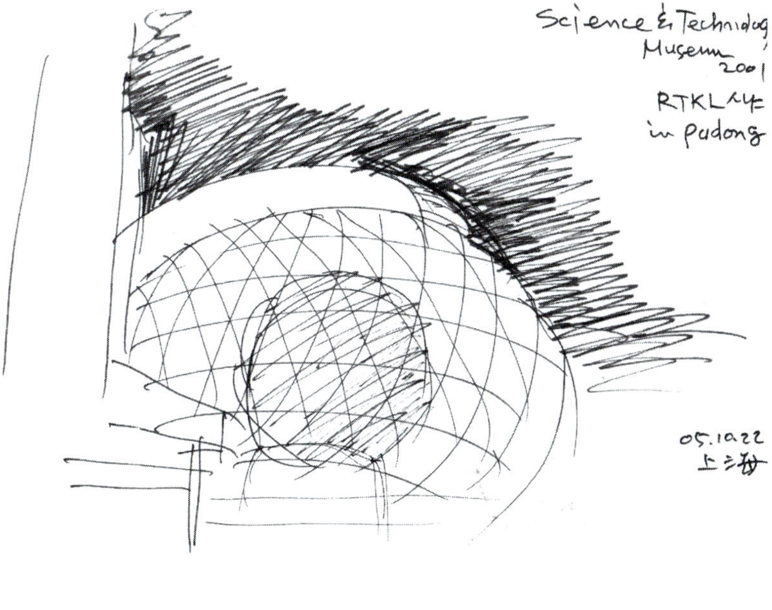

아파트 전경

Complete view of Apartment Shanghai

상하이 아파트들은 옥탑과 지붕디자인을 살려 전체적 스카이라인이 독특하게 보이도록 하는 특징이 있다. 최근 기능주의의 모더니즘이 점차 아파트에서 자취를 감추고 있으며 평면의 다양화 뿐 아니라 입면에서도 다양한 디자인을 볼 수 있다.

Apartments of Shanghai have the characteristics of unique skyline by improving designs of rooftop and rooftop rooms. recently, modernism focusing on functions is gradually disappearing from apartment buildings and now can find variety of floor plans as well as erection design.

서커스장

Shanghai Circus City

상하이서커스성은 꿍허신루(共和新路)에 있다. 면적 2.25ha로 상하이시 북구의 문화, 체육 오락 센터이다. 이곳은 서커스장, 리허설실, 동물방, 오락실, 배우대기실의 5개 부분으로 조성되었다.
상하이 서커스는 북경의 서커스보다 더 화려하고 고난이도의 묘기가 속출하는 상하이의 볼거리로 널리 알려져 있다.

Shanghai circus palace is located at Kungheosinru. The area is 2.25ha and is the culture and sport entertainment center in the north district of Shanghai city. This place consists of 5 sections of circus link, rehersal room, animal room, entertainment room, and performers waiting room.
Shanghai circus is more splendid than Beijing circus and has more difficult and sophisticated skills to attract tourists in Shanghai.

심문고관 전경

Complete view of Simmungogwan Shenyang

중국 심양에 위치한 전통 건물로 도시속의 오아시스처럼 보인다.
It is located in Shenyang of China and as a traditional building, it looks like an oasis in city.

The Great Wall of China

총 연장은 약 2,700km로 중국의 거리 단위로 환산하면 1만리를 넘기 때문에 '만리장성' 이라는 이름이 붙었다. 진시황이 흉노족의 침입에 대비해 10여 년간 쌓았으며, 현재의 모습은 명나라 때 완성되었다. 지금은 관광객을 위한 다양한 위락 시설과 엘리베이터 등이 찾는 사람들의 편의와 즐거움을 함께 주고 있다.

The total extension length is about 2,700km and when it is converted in Chinese distance unit, it is more than 10,000 Ri and was named as '10,000 Ri wall.' Qin Shi Huang had built over 10 years to prevent attacks of Xiongnu tribe and the current feature was completed during Ming dynasty. Now it provides various entertainment facility and elevators for the comfort and pleasure of tourists.

타이완은 원래는 부속제도인 펑후제도[澎湖諸島], 훠사오섬[火燒島], 란위섬[蘭嶼] 등 79개의 도서를 합하여 중국의 1개 성(省)인 타이완성을 이루었다. 1949년 이래 타이베이[臺北]를 임시수도로 정하고 있는 타이완 국민정부의 사실상의 지배지역은 타이완 및 푸젠성에 속하는 진먼섬[金門島]과 마쭈섬[馬祖島]이다. 한편 지방정부로서의 타이완성 정부는 타이중[臺中] 교외의 중싱신춘[中興新村]에 있다.

THAIWAN

Taiwan originally consisted of 79 islands including Penghu(Pescadores) islands, Green isalnd, and Rwanwi island to become one province of China as Taiwan province. Since 1949, Taiwanese people' s government with a temporary capital of Taipei had actual control over Taiwan and Jinmon Island and Matus island belonging to Fuzen province. In addition, as a local government of Taiwanese provincial government, it is located at Jungshingsinchun in the suburbs of Taijung.

타이베이

한민족(漢民族)이 이곳으로 이주하기 시작한 것은 17세기 말부터이며, 18세기 초에는 단수이 강 연안에 마을이 형성되었다. 이 마을이 완화[萬華]이다. 이곳은 하항(河港)으로서, 오지에 사는 고산족(高山族)과의 교역기지였으며 대륙과의 교역을 위한 주요 무역항으로 번창하였다.

Taibei

It was from the end of the 17th century when Han tribe had moved to this place and in early 18th century, they formed villages near the riverside of River Dansui. This village was Wanhwa. This was a river port and was a trading place with people living far inland of high mountain and had prospered as an important trading post with the continent of mainland.

타이베이 금융센터

Taipei Financial Center Taipei

높이 508m에 지상 101층, 지하 5층 건물로 말레이시아의 페트로나스타워(451m)보다 키가 50m 이상 큰 '세계 최고층 빌딩' 이다. 대만의 세계적 건축가 리쭈웬(李祖原)이 설계한 이 빌딩은 '8' 을 좋아하는 중국인들의 의식을 담고 있다. 8층씩 묶어 8개의 층으로 올려 그런 모양을 만들어 냈는데, 숫자 '8' 이 중화문화에서는 '성장, 번영, 발전' 을 의미하는 한자 '發' 과 발음이 같은 '길한 숫자' 이기 때문이다. 이런 의미를 담아 전체 건물은 이 꽃이 포개져서 하늘로 올라가는 모습을 형상하고 있다. 이렇듯 타이베이 101빌딩은 외양에서부터 중화적인 색채와 사상, 그리고 세계로 뻗어 나가는 국제도시 타이베이의 모습이 물씬 풍겨난다. 주변은 세계무역센터의 전시장들과 국제컨벤션센터(國際會議中心), 세트라 빌딩이 자리잡고 있는 타이베이 경제중심지로 매년 수많은 교역 전시회와 각종 박람회가 개최되고 있다.

This is 'the highest building in the world' with the height of 508m in 101 floors above the ground and 5 stories of underground and is higher over 50m than the Petronas Tower in Malaysia (451m). The building designed by the world famous architect, Lee, Chu-Wen and contains the consciousness of Chinese who like the number '8.' It tied in bundle of 8 stories in 8 different tiers and it is because the number '8' is a 'lucky number' that has the same pronunciation of a Chinese character of 'Bal' that means 'growth, prosperity, and development' in Chinese culture. The whole building expresses the shape of flowers folded together to fly to the sky. Like this, 101 building in Taipei depicts the shape of an international metropolitan city of Taipei that extends to the world with its color and idea of the center of the world in its exterior. In the vicinity, there are exhibition center of the world trade center, international convention center, and Cetra building as the central point of Taipei economy and numerous trade conventions and expos are held here every year.

정식명칭은 인도공화국이다. 힌디어로는 바라트라고 한다. 수도는 뉴델리이다. 아시아 문명의 원천으로 불교가 발상한 곳이며, 천축(天竺)이란 이름으로 예부터 알려진 곳이다. 1947년 8월 15일 3세기 반에 걸친 영국의 지배를 벗어나 독립한 민주국가로, 영국연방에 속한 나라이다. 국토면적은 세계 7위이며, 인구는 세계 2위이다.

Official name is the Republic of India. In Hindi, it is called as Bharat. The capital is New Delhi and as original place for Asian civilization, Buddhism was born in here and it was known as the name of Cheonchuk from the past. It became independent on August 15, 1947 as a democratic country from Great Britain after 3 century and belongs to the Commonwealth of Great Britain. The national territory is the 7th largest in the world and population is the second in the world.

INDIA

아메다바드
인구 약 89만 9천(1991). 야무나강(江) 우안에 있는 지방행정의 중심지이다. 델리 남동쪽 200km 지점에 있다. 동쪽에는 갠지스강 유역의 광대한 평야가 전개되고, 북쪽은 야무나강 연안을 따라 델리를 거쳐 펀자브 지방의 평야에 연속되어 있다.

아그라
무굴 제국이 수도를 델리로 옮길 때까지 1564~1658년 약 1세기 동안 수도로서 북부 인도를 지배했다. 영국 동인도회사의 지배를 받을 때도 주변지구의 행정중심지로서 중시되었으며 오늘날도 지방행정부의 청사가 있다.

천디가르
인도와 파키스탄의 분리 독립으로 인도 쪽에서 펀자브주의 새로운 주도(州都)가 필요해졌기 때문에 건설되었으며, 1953년에 펀자브 주정부가 심라에서 이곳으로 이전하였다.

델리
갠지스강(江)의 지류인 야무나강의 서쪽 기슭에 있으며, 펀자브 지방과 갠지스강 유역과의 교통 중심지여서 고대부터 이 지방의 정치 · 문화 · 경제의 중심을 이루었다.

자이푸르
도로와 철도 등 교통망이 정비된 상공업 중심지이다. 1727년 자이푸르 번왕국(藩王國) 왕 마하라자 자이싱 2세가 건설하였으며, 자이푸르는 '자이왕의 성'이라는 뜻이다.

Ahmedabad
Population is about 899 thousand people. (In 1991) and is the center of local administration located in the right side of River Yamuna. It is located about 200kmin the southwest of Delhi. In the east, a huge plain is stretched in the region of River Genzis and in the north, it is connected to plain of Punzab region alongside River Yamuna via Delhi.

Agra
Until Empire of Mugul moved the capital to Delhi, it was the the capital for about 1 centurt between 1564~1658 and controlled the northern India. During the governance of British East Company, it was the center of administration of the region and there is local government building in these days.

Chandigarh
It was built as a new capital was needed in the Punzab province of India after the separation independence of India and Pakistan and in 1953, provincial governmen 샐 Punzab moved to here from Simra.

Delhi
It is located in the west foothill of River Yamuna that is a branch stream of River Genzis and as the center of traffics between Punzab region and River Genzis region , it was the center of politics, culture, and economy from the old period.

Jaipur
It is central commercial and industrial city with well established transportation network of roads and railroads. In 1727, it was built by King, Maharaja Jaijing II of Bun Kingdom of Jaipur and Jaipur means the 'castle of King, Jai.'

라니카 하지라

Ranika Hazira Ahmedabad

자미 마스지드

Ahmedabad Jami Masjid

이 도시의 건립자 아흐메드 샤에 의해 1424년에 세워진 큰 규모의 회교 사원이다. 올드 델리에 있으며 1644년 착공 내부는 2만명이 한꺼번에 예배볼 수 있는 넓은 마당과 3개의 큰 문, 본당, 높이 40m의 전망탑으로 구성되었다.

This is a large Islam mosque built in 1424 by Ahmed who was the founder of the city. It is located in old Delhi and was started to be constructed in 1644 and the inside consists of a huge ground where 20,000 people can have prayer service at the same time, 3 large gates, main hall, and 40m high observatory tower.

계단우물

Dada Hari War Ahmedabad

4각형에서 8각형으로 이어지는 조형과 공간의 기하학적 구성은 전 세계 어느 곳에서도 찾아볼 수 없는 독창적 우물이다. 1499년 건조되었으며 지하레벨로 들어가는 통로를 따라 지상에서 들어오는 빛은 가히 환상적이다.

This is a very unique well with geometric composition of model and shape that connects from square to octagonal shape, which cannot be found anywhere else in the world. It was built in 1499 and sunlights that come through tunnel going into the basement is really fantastic.

인도경영대학원

Ahmedabad Indian Institute of Management

루이스 칸의 작품으로 1962년에 착공되어 1947년에 완공되었다. 루이스 칸이 이 건물의 공사관계로 인도 출장을 다녀오는 길에 1974년 3월 17일 뉴욕의 기차역에서 심장마비로 사망했다. 건축가는 세상을 떠났으나 그가 죽은 후 이 작품은 더욱 빛났다. 펜실베니아 예술아카데미의 퍼니스상, 1979년 미국 건축가 협회 25주년상, 1989년 아가칸(Aga Kahn)상을 수상하였다.

This is the work of Louis Kahn and was started to be constructed in 1962 and was completed in 1974. Louis Kahn passed away at New York railroad station by heart attack on March 17, 1974 on his way back from the travel to India because of the construction of this building. Architect left this world but his work became more famous. The building had received the Furness award of Pennsylvania Arts Academy, the 25th anniversary award of American Architect Association in 1979, and Aga Kahn award in 1989.

미나버자르

Meena Bazaar Ahmedabad

Agra Sikandra

아그라로부터 12km 정도 떨어진 시칸드라에는 악바르 대제의 묘가 있는데 악바르 양식이라고 불리는 붉은 사암과 대리석으로 만들어진 독특한 양식의 건물이다.
성채를 두르고 사방으로는 붉은 사암으로 4개의 문을 만들었는데 이 문은 각각 힌두교, 회교, 기독교 그리고 악바르 스스로 창안한 종교에 따른 양식에 의해 지어졌고, 이 문을 통해 들어가면 만나게 되는 큰 규모의 정원 한가운데에 악바르의 무덤이 있다. 내부로 들어가면 사방으로 커다란 회랑이 높이 위치해 있고, 중앙의 컴컴한 복도를 따라 들어가면 신비스런 분위기를 풍기는 악바르의 석관이 자리하고 있다.
이 무덤은 악바르 자신이 생전에 건축을 시작하였으나 1605년 아들 제항기르에 의하여 완공되었다.

In Sikandra that is about 12km away from Agra, there is a tomb of The Great Emperor, Akbar and this is a unique building made of red sand rock and marbles in Akbar style.
It is surrounded by castle walls and 4 gates made of red color sand rock were built according to the style of Hinduism, Mohammedanism, Christianity, and the religion Akbar had created. Entering through these gates, there is a tomb of Akbar in the middle of a huge garden. Entering into the tomb, in all four directions, large corridors are located in high positions and following the central dark corridor in the middle, stone coffin of Akbar is found with mystic mood.
This tomb was started to be constructed while Akbar was alive and was complted in 1605 by his son, Jahanggir.

시칸드라

Sikandra Agra

악바르성 게이트

Agra Akbar´s Tom Gate

악바르대제의 무덤으로 들어가는 입구문에서 성을 보면, 프레임으로 들어오는 한편의 수채화를 감상하는 듯하다.

When we look at the castle from the entrance gate entering into the tomb of The Great Emperor, Akbar, it feels like watching a piece of watercolor painting in a frame.

타지마할

Taj Mahal Agra

인도 무굴제국의 황제 샤자한의 2번째 아내 뭄타즈 왕비를 위해 지어진 건물이다. 황제의 너무 깊은 사랑으로 항상 전쟁터에까지 데리고 다녔으나 마지막으로 15번째 아이를 낳다 숨을 거두었다.

This is a building built for the second wife, Queen, Mumtaz by Emperor of Indian Mugul Empire, Shajahan. The Queen always was with the Emperor in all wars as they were in such deep love but the Queen passed away while giving a birth to the 15th child.

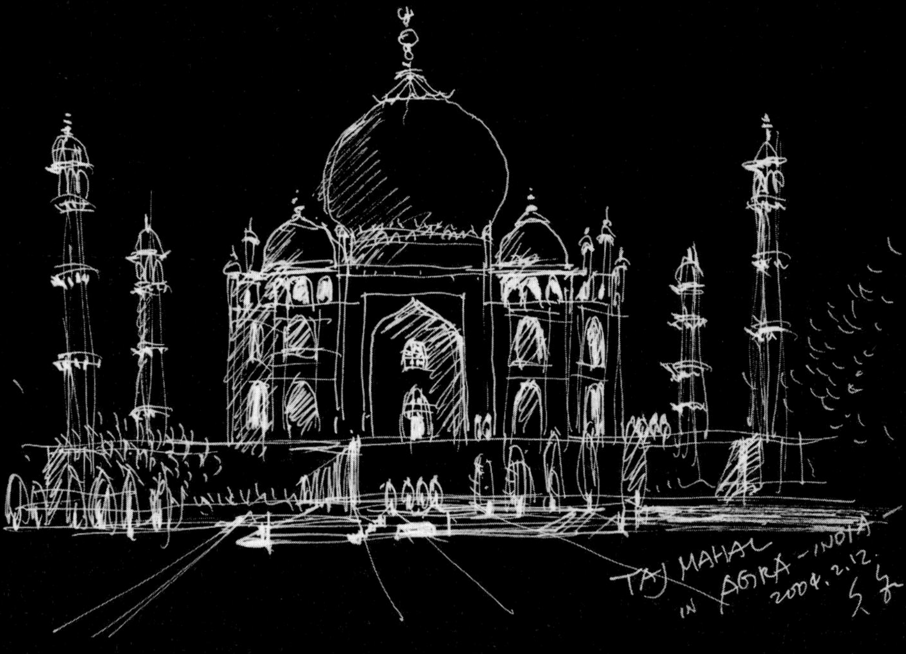
TAJ MAHAL
IN AGRA - INDIA
2004. 2. 12.

고등법원

Court House(Le Corbusier) Chandigarh

찬디가르 도시계획의 일환으로 르 꼬르뷔제가 디자인한 법원 건물이다. 건물 뒷쪽의 램프를 따라 위로 오를 수 있으며 지붕 밑에서는 히말라야 기슭까지 볼 수 있는 테라스와 펜트하우스가 있다. 전면부는 뜨거운 태양을 막기 위한 우산 모양의 아치가 계획되었으며 좌측 기둥과 벽에는 르 꼬르뷔제 특유의 원색으로 도색되어 있다.

This is a court building designed by Le Corbusier as a part of city planning of Chandigar. Through the lamp in the backside of the building, people can climb up the building and there are terrace and penthouse under the roof where people can observe the foothill of Himalaya. The front side of the building is designed in arch style like an umbrella to prevent sunlight and the left side pillar and wall is painted with unique original color of Le Corbusier.

High Court BLD'G
Le Corbusier

오픈핸드

Open Hand(Le Corbusier) Chandigarh

찬디가르에서 가장 상징적인 높이 16m의 조각품이다. 전면에서 보면 한편으로 손을 펼치고 있는 모습이, 또 다른 한편으로는 새의 모습이 이미지로 다가온다. 이 기념비는 인간과 자연이 함께 공생하고 있음을 표현하고 있다.

This is a 16m high sculpture as the most symbolic artifact in Chandigar. If looked from the front, it is the shape of open hand or as the image of flying bird. This monumental tower expresses the coexistence of the nature and human being.

정부청사

Chandigarh (Le Corbusier)Secretariat

전면의 기하학적 구성으로 된 수많은 프레임과 창은 모듈러와 황금비로 된 르 꼬르뷔제만의 한편의 독창적 그림이다. 내부는 좌우로 램프가 있으며, 르 꼬르뷔제의 근대 건축 5원칙에 충실한 정원이 있다.

So many frames and windows composed in geometric structure in the front is a unique picture made of modular and extreme and mean golden ratio by Le Corbusier. Inside, there are lamps on both sides and a garden fulfilling the 5 principles of modern architecture of Le Corbusier.

주의회당

Assembly Building(Le Corbusier) Chandigarh

르 꼬르뷔제가 아메다바드 상공을 비행기로 지나다 강가에 늘어선 쿨링타워에서 아이디어를 얻어 설계했다. 따라서 의회당의 모습은 마치 원자로, 여객선의 굴뚝같은 원통형과 입방체가 커다란 매스위에 얹혀 있는 듯한 모습이다.

Le Corbusier had designed by obtaining idea from cooling towers erected alongside river while flying over the sky of Amedabad. Therefore, the shape of capitol is just like a nuclear reactor and chimney of passenger ship as circular cubics are put on a huge mass.

Chandigarh Architecture University (Le Corbusier)

이 건축 미술 대학은 멀리서도 금방 알 수 있듯이 르 꼬르뷔제의 '모듈러'가 전면 입구에 그려져 있으며 뜨거운 햇빛을 막기 위한 차단 그릴이 병풍처럼 둘러져 있다. 건물 전체는 정방형의 마당(코트)을 감싸고 있는 사각형으로 되어 있으며, 그 주위로 화랑과 교실들이 배치되어있다.

This architecture arts university can be recognized easily even from far away as 'Modular' of Le Corbusier is painted in the front entrance and interception grill is surrounded as a screen to prevent hot sunlights. The whole building is made in regular square shape surrounding the ground (court) and corridors and class rooms are arranged.

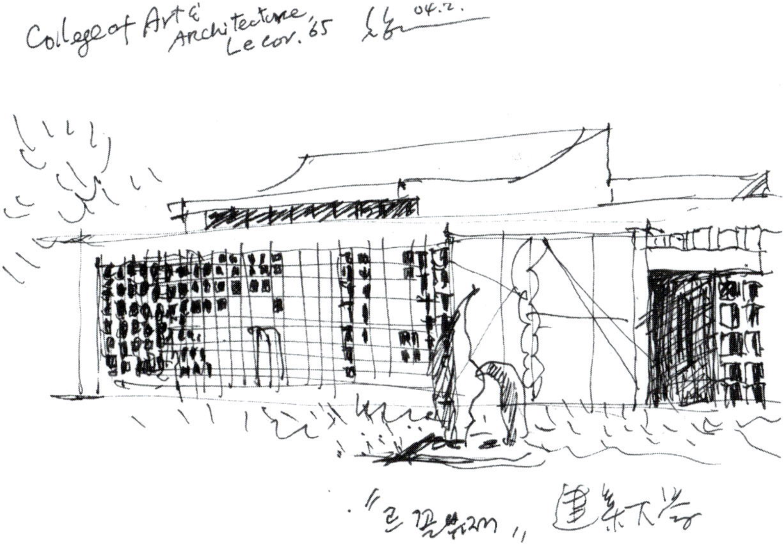

붉은성

Red Fort Delhi

올드델리의 랜드마크라 할 정도로 규모와 웅장함에서 독보적인 유산으로 생각되어지는 붉은 성은 타지마할을 건축하기도한 무굴 황제이자 건축광이었던 샤자한이 1639-1648년에 걸쳐 지은 성으로 지금의 올드델리가 수도였던 시절에 왕궁으로 사용했던 건물이다. 1857년 세포이항쟁으로 성의 대부분이 파괴되어서 관광객들은 그전의 실제모습을 볼 수는 없다. 1857년 전까지만 해도 이곳은 미니도시라 생각되어질 정도였고, 또한 이곳에는 3,000여명의 사람들이 살았었다고 한다.

This red castle as a landmark of old Delhi is unique in its size and grandeur was built during 1639-1648 by Mugul Emperor and an avid architecture mania, Shajahan who also built Taji Mahal and was used as royal palace when old Delhi was the capital. Since most of castle was destroyed in 1857 during Sepoi protest, tourists can not see the original feature in the past. Until before 1857, this place was considered as a mini city and in addition, over 3,000 people had lived there.

릭샤

Delhi Rickshaw

사이클 릭샤는 자전거 뒤에 사람이 앉을 수 있는 바퀴 달린 의자를 붙여놓은 3륜차로서, 다른 교통수단에 비해 단거리 이동을 할 때 주로 이용한다. 도심을 질주하는 오토 릭샤는 동남아시아에서는 흔히 툭툭이라고도 부르며, 인도에서는 인디언 헬리콥터라고도 한다. 릭샤 내부에는 거리별 요금을 계산하는 장치가 달려있으나 거의 이용되는 경우가 없으며, 승차전 릭샤왈라와 흥정하여 결정한다.

Cycle rickshaw is a tri-wheel chair attached to the behind of a bicycle and is mainly used for traveling a short distance. Automatic rickshaw that runs across the cities is often called as Tuk Tuk in the Southeast Asia and is called as an Indian hellicopter in India. Inside Rickshaw, a device calculating fare is attached but is rarely used and therefore, price is decided after negotiation with Rickshawaller before riding.

라지가트

Raj Ghat Delhi

마하트마 간디 거리에 있는 화장터로 현재 추모공원으로 조성되어 있다. 검은 대리석의 대좌 정면에는 간디의 마지막 말 헤이람(Hai Ram)이 새겨져 있다. 이 곳은 인도의 자유와 민주주의를 상징하는 성역의 공간으로 인도인들에게는 평생 한 번은 가봐야 할 곳으로 인식되어 있다.

As a crematory located on the Mahatma Gandhi street, it is now landscaped as a memorial park. In the middle of black marble in the front of facing seat, the last word of Mahatma Gandhi, Hai Ram is engraved. This is a sacred place symbolizing the freedom and democracy of India and is considered as a must place to visit for all Indian people during their life time.

쿠트브 미나르

Delhi Qutb Minar

터키계의 이슬람교도로서 노예왕조를 창시한 쿠트브 웃 딘 아이바크가 인도 북부를 점령한 뒤 13세기에 건설하였다. 이 탑은 아프가니스탄에 있는 잠무탑을 본뜬 것으로서 이슬람의 승리를 기념하기 위하여 세운 인도에서 가장 높은 석조탑으로 높이가 72.5m에 이른다. 내부에는 379단의 나선 모양 계단을 설치하여 각 발코니로 통하도록 하였다. 1993년 유네스코(UNESCO:국제연합교육과학문화기구)에서 세계문화유산으로 지정하였다.

It was built in the 13th century by Kutub Ud Din Aibak who had created slave kingdom as a Turkish Islamic after conquering the northern India. This tower imitated Jamu Tower in Afghanistan and is the tallest stone tower built to commemorate the victory of Islam and is 72.5m high. Inside the tower, 379 circular steps are installed to lead to each balcony. In 1993, UNESCO has designated it as the world cultural heritage.

마제스틱 자이푸르

Majestic Jaipur Jaipur

델리에서 남서쪽으로 약 266km 떨어진 자이푸르는 광대한 타르사막을 끼고 있는 라자스탄의 주도이다. 1728년 이 지방에서 세력을 떨쳤던 사와이자이싱 2세(1699~1744)가 만들었다. 도시의 이름조차 왕의 이름을 본떠 자이푸르('승리의 도시'라는 뜻)로 명명됐으며 푸르는 성벽으로 둘러싸인 도시라는 뜻. 이름처럼 구 시가지는 7개의 문을 지닌 성벽이 병풍처럼 둘러싸고 있다. 자이푸르의 암베르성은 무갈 제국과 라자스탄 양식을 도입하여 무척 아름다운 모양으로 건축되었으며, 내부에 있는 접견실은 붉은 사람으로 조각된 기둥과 벽이 아름다우며, 더욱이 바닥의 흰 대리석은 저녁 달빛이 실내로 흘러들어오게 하여 방 전체를 달빛으로 채워, 보는 사람들을 황홀하게 만든다.

Jaipur that is 266km away in the southwest of Delhi is the capital of Lajastan with a huge Tar desert. It was built in 1728 by King, Sawai Jai Singh II (1699~1744) who controlled the region. The name of city was named as Jaipur (Meaning 'city of victory') after the name of King and pur menas city surrounded by castle. Like it name, the old section of the city is surrounded by castle walls with 7 gates like a floding screen. Amber castle of Jaipur was built in the styles of Mugal Empire and Lazastan style and is very beautiful. Receiving room inside has beautiful column and walls made of red sculptures of human and in particular, white marbles on the floor attract moonlights and fill the room and it makes people fascinated.

MAJESTIC
JAIPUR
- INDIA -
04/2/12

PHILIPPINES

면적은 30만 76㎢, 인구는 8116만 1000명(2003)이다. 인구밀도는 270.5명/㎢(2003)이다. 정식명칭은 필리핀공화국(Republika ng Pilipinas)이다. 수도는 마닐라이고 공용어로 영어와 타갈로그어를 사용하고 있다. 필리핀해(海), 셀레베스해, 남중국해의 경계를 이루며 타이완[臺灣]섬과 보르네오섬, 셀레베스섬 사이에 있다. 필리핀이라는 국명은 16세기 중엽 이 지역으로 파견된 에스파냐 탐험가 빌리아로보스가 당시의 에스파냐 황태자 필립의 이름을 따서 붙인 것이다.

The area is 300 thousand 76㎢ and population is 81 million 161 thousand people. (In 2003) Population density is 270.5 people/㎢. (In 2003) Official name is the Republikang Pilipinas. The capital is Manila and use English and Tagalog as official language. It is bordered with Philippine ocean, Celebes Sea, and the south Chinese ocean and is located between Taiwanese island, Borneo Island, and Celebes Island, . The name of Philippine was named by a Spanish explorer, Villalobos who was dispatched to this region in the middle of 16th century by quoting the name of prince of Spain, Philip at that time .

마닐라
세계에서 가장 좋은 항만으로 일컬어지는 마닐라만(灣)에 임한 항구도시로, 시가지는 파시그강(江)을 끼고 그 남북으로 펼쳐진다. 북쪽에 비옥한 중부 루손 평야를, 남쪽에 남부 루손의 화산성 저지를 끼고 있다.

Manila
It is a port city facing with Manila Bay known as the best harbor in the world and the city runs across in the north and the south with River Pasig. In the north, there is fertile Luzon plain and in the south, there is lower land of Luzon Volcano.

필리핀 문화센터

Manila Cultural Center of the Philippines

로하스 대로의 중앙은행 건너편, 마닐라 만 쪽에 있는 현대적인 빌딩. 곡선과 직선의 상하 배합으로 특이한 인상을 주는 건물이다. 내부에는 회화전시실과 공회당, 소극장, 미술관 등이 있다. 공연이 펼쳐지는 저녁 해가 지는 노을에 비춰진 건물은 너무나 아름답다.

It is a modern building located near Manila Bay in the opposite of Central Bank on Lohas Avenue and provides a very unique impression in combination of straight and curve lines. Inside, there are pictures exhibition hall, public gathering place, small theater, and art museum. The building with performance in the evening with the reflection of sunset is so beautiful.

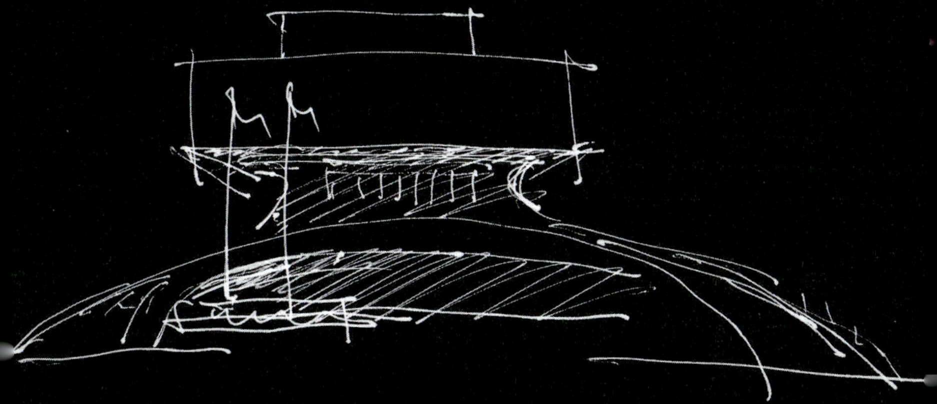

TURKEY

동양과 서양에 걸쳐진 나라 터키의 정식명칭은 터키공화국이다. 수도는 앙카라이며, 공용어로 터키어를 사용하고 있다. 동쪽으로 이란 · 아르메니아 · 그루지야, 남쪽으로 이라크 · 시리아, 북서쪽으로 불가리아 · 그리스에 접하고, 북쪽으로 흑해, 남쪽으로 지중해, 서쪽으로 에게해(海)에 면한다. 소아시아(아나톨리아) 반도 전부, 보스포루스해협 · 다르다넬스해협 · 마르마라해를 사이에 두고 유럽의 발칸 반도, 동(東)트라키아 지방에 걸쳐 있다.

The Official name of Turkey that is a country that connects the Orient with the West is the Republic of Turkey. The capital is Ankara and people use Turkish as an official language. In the east, it is bordered with Iran, Armenia, Gruzia and in the south, with Iraq and Syria, in the northwest with Bulgaria and Greece, in the north, with Black sea, in the south with Mediterranean , and in the west with Aegean ocean. It is stretched over the whole small Asia (Anatolia) peninsula, Balkan peninsula and the East Thracia region in between Bosporus strait , Dardanelles strait, and Sea of Marmara.

이스탄불
트로이
카파도키아
파무칼레
보두룸

보드룸

Bodrum

이스탄불에서 남쪽으로 600km 떨어진 곳에 위치해 있는 항구 도시로 유럽인들의 휴양지로, 그리스 섬과의 연결지로 인기를 끌고 있다. 고대7대 불가사의 중 하나인 마우소루스의 묘가 있고 고대 그리스의 역사가 헤로도투스(Herodotus)의 고향으로도 알려져 있다. 1년 사계절 맑은 날씨를 유지하는 에게해 최고의 리조트 도시로 하얀 집들, 푸른 바다위에 떠있는 요트들이 관광객을 끄는 곳이다.

It is a port city located 600km away in the south from Istanbul and is a resort for European and is also popular as the connection place with island of Greece. Tomb of Mausolus that is one of the 7 wonders of the world is located here and it is the hometown of a Greek historian, Herodotus. This is the best resort town in Aegean ocean with clear weather in all four seasons of a year and white houses and yachts floated in the sea attract tourists.

하얀 집들
지중해
보드룸 - Turkey
03.1.22

보드룸성

Castle Bodrum Bodrum

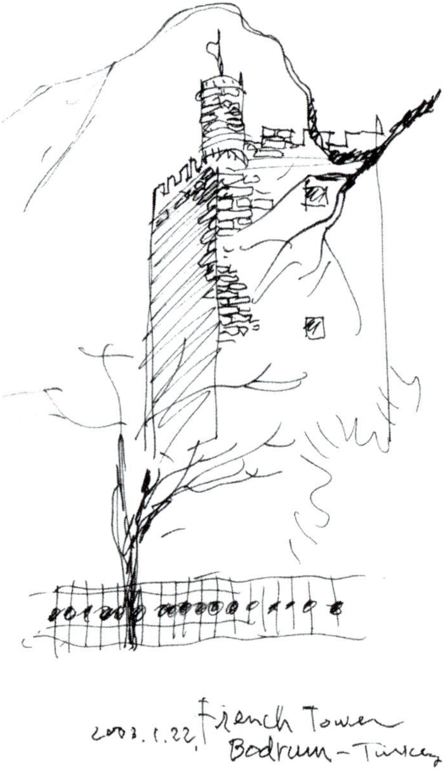

15세기 십자군에 의해 건설되었으며 해방자 St. Peter의 성으로 더 알려져 있는데 이 일대는 서해안의 기독교도들이 십자군 시대 동안 모여 지낸 곳으로 1세기 이상 기독교 사회를 이루며 살았다. 1523년 터키의 통치하에 들며 이곳은 군사방어요새, 감옥 그리고 공공목욕탕 등 여러 용도로 사용되었는데 지금은 수중 박물관(Underwater Archaeology Museum)으로 사용되고 있다. 계단을 통해 성의 입구로 들어가면 여러 세기에 걸친 고대 항아리(두 손잡이) 가 전시되어 있고 장인의 작업 모습도 볼 수 있다.

반대편 건물에는 7세기의 동로마 배의 선미를 제작해 전시하고 있는데 계단을 올라 선미에 오를 수도 있고 내부의 모습도 관찰할 수 있다.

It was built by the Crusade in the 15th century and well better known as the castle of a liberator, St. Peter and this is the region where christian in the west coast lived here as the society of christians for over one century during the period of Crusade. In 1523, it fell under the control of Turkey and was used as diverse purposes such as military defence fort, prison, and public bath and is now used as Underwater Archaeology Museum. When entering into the castle through stairs, old potteries (two handles) over several centuries are exhibited and we can see craftsmen work.

In opposite building, sterns of the east Roman ships during the 7th century are displayed and people can go on board through stairs and can take a look inside of the ship.

Harbour Tower — CASTLE
03.1.23 BODRUM-T

카파도키아

Cappadocia

이 곳은 4세기경 로마제국의 그리스도교 박해를 피해온 그리스도교도들이 만들기 시작했다. 7세기 후반부터 이슬람의 침입을 피해온 그리스도교도의 이주가 늘어나면서 11세기 초에 이르러서는 인구가 7만명, 암굴 성당과 수도원 수는 한때 360개를 넘어섰다고 한다. 인구가 늘자 카이마클루 지구 같은 곳에는 굴 안에 주거지, 학교, 식료품 저장소, 우물, 환기용 굴뚝, 묘지 등을 갖춘 지하도시도 건설되었다. 프레스코화는 8~9세기 전반의 우상파괴운동 때 파괴되어 초기의 것은 거의 남아 있지 않으며, 9세기 후반~13세기까지의 작품들이 대부분이다. 비잔틴 양식이 많은데, 작풍은 대부분 소박하고 검소하다.
엘말루 킬리세 성당(사과 성당)과 토칼루 킬리세 성당(버클성당) 카랑루크 킬리세 성당(어둠의 성당), 메리에마나 길리세(성모마리아 성당) 차루클루 킬리세 성당(샌들 성당)등이 유명하다.

This place was started to be built by christians who suffered from Roman Empire in circa 4 century. In the later part of the 7th century, christians' relocation who escaped from Islamic attacks had been increased and in the early 11th century, population was 70,000 people and the number of cave churches and monastery was over 360 at one time. As population had been increased, underground city that included residence, school, food storage, spring well, chimney for ventilation, and tomb was established in Kaymakli district. Presco pictures were all destroyed because of idol destruction movement in the 8th~9th century and the early pictures did not survive and most of them were during the later part of the 9th century and the 13th century. Most of them are Byzantine style and mood of works is simple and frugal.

Elmali Kilise church(Apple Church), Tocali Kilise church(Buckle Church), Karanlik Kilise church, Meryemana Kilise church and Carikli Kilise church (Sandal Church) are famous.

성소피아 성당

St. Sophia Cathedral (Ayasofya) Istanbul

동로마 제국이 이 곳으로 옮기고 그리스도를 위해 기념비적인 건물을 지었다. 자장 아름다운 건축의 특징은 중앙의 지름 33m의 돔이 주변의 작은 창에 의해 마치 하늘에 떠있는 듯한 황홀함을 주는 것이다. 내부는 비잔틴의 특징인 모자이크로 그려진 벽화가 아름답게 펼쳐져 있어 찾는 사람의 주목을 끌고 있다.

East Roman Empire had relocated to this place and had built a monumental building for Jesus Christ. The most beautiful characteristics of the architecture is the central dome with the diameter of 33m and it is fascinating that it looks like floating in the sky by many surrounding small windows. The interior is decorated with mosaic wall paintings in Byzantine characteristics to attract people.

Istanbul Sultan Ahmet Mosque

성 소피아 사원 바로 건너편에서 볼 수 있는데 블루모스크로 널리 알려져 있는 이 사원을 터키 사람들은 술탄 아흐멧 모스크라고 부른다. 오스만 터키 제국의 14번째 왕인 아흐멧 1세가 20세 때 성소피아 사원을 모방하여 세운 회교 사원으로 이스탄불에서 가장 큰 사원이다. 아흐멧 1세는 당시 유명한 건축가였던 시난의 제자 메흐멧 아아(Mehmet Aga)에게 명하여 1609년 건축을 시작하였고, 7년 후 1616년에 완성하였다. 오스만 제국의 왕들은 종교적인 중요한 선언을 이곳에서 가졌다. 종교 휴일에는 이곳에서 축제를 열고, 메카로 순례를 떠날 때는 이곳이 출발지가 되었다.

Turkey people call this mosque that is located right in opposite of St. Sophia temple and is known as a blue mosque as Sultan Ahmet Mosque. This is an Islamic mosque built by Ahmet I who was the 14th king of Ottoman Turkey Empire in order to imitate St. Sophia mosque when he was 20 years old and is the largest mosque in Istanbul. King Ahmet I had ordered Mehmet Aga who was a student of the famous architect of that time, Sinan to build the mosque and he began to build in 1609 and completed in 7 years in 1616. Kings of Ottoman Empire had declared important religious statements in here. During religious holidays, feasts are held here and this was the starting point for a pilgrimage to Mecca.

돌마바흐체 궁전

Dolmabahce Mosque Istanbul

19세기 술탄 압둘메지드가 유럽의 라이벌들과 경쟁하기 위해 지어졌다는 "돌마바흐체 궁전"은 14톤의 금과 40톤의 은으로 장식되었으며, 궁전내부는 진귀한 카펫과 장식품들로 채워져 있다. 건축적 가치는 그리 높지 않으나 해안의 아름다운 풍경과 어울려 현재 많은 관광객들에게는 최고의 관광 명소가 되어버렸다.

"Dolmabahce Mosque" that was built by Sultan Abdulmecid in the 19th century in order to compete with European rivals is decorated with 14 tons of gold and 40 tons of silver and palace interior is filled with rare carpets and decorative artifacts. Architectural value is not that high but it is now one of the best tourism spot for many tourists together with the harmony with beautiful coastal sceneries.

Dolmabahce
Mosque
2003.1.26.
유람선에서

슐레이마니예 사원

Sleymaniye Temple Istanbul

슐레이마니예 사원은 이스탄불에서 두 번째로 큰 사원으로 오토만 왕족의 거주지였던 오토만 올드 팔레스 자리에 세워졌다. 이 사원은 200×140m에 달하는 정원의 한 가운데에 세워졌고 앞에는 주랑으로 된 정원이 있으며, 뒤에는 술레이만과 그의 부인인 휴렘 술탄의 무덤이 있는 커다란 묘지가 있다. 이 정원을 중심으로 사원의 부속 건물들이 주위에 있다.

Sleymaniye Temple is the second largest mosque in Istanbul and was built in the place of the old Ottoman Palace where was the residence of Ottoman royal family. The mosque was built in the middle of a garden with the size of 200 × 140m and there is a garden consisted of main columns in the front and there are huge tombs of Sleymaniye and his wife. Auxilliary buildings of the mosque were built surrounding this garden.

히에라포리스의 원형극장

Pamukale Arena theater in Hierapolis

히에라폴리스의 가장 대표적인 유적 중의 하나로 15,000명을 수용할 수 있는 거대한 야외 원형극장이다. 2세기 하드리안 황제 때 건설되고 3세기 셉티무스 세베루스 황제 시대에 개축되었고 현재 복구 공사가 진행되고 있다. 중앙 무대 주변의 대리석 조각이 아름답다. 중앙에는 대리석으로 된 무대와 조각이 있으며 이곳에서는 주로 디오니수스의 일생을 자주 공연했다.

One of representative ruins in Hierapolis is this huge outdoor arena theater that can accommodate 15,000 people. It was built during Hadrian emperor in the second century and was rebuilt during Seberus emperor in the third century and is now under restoration construction works. A marble sculpture in the central stage is beautiful. There are stage and sculpture made of marble in the middle and in here, performance of the life of Dionysos was mainly held frequently.

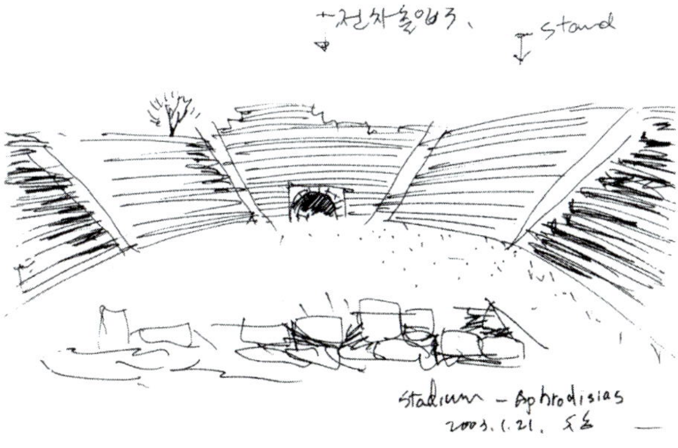

랍세키 마을 전경

Complete view of Lapseki Village Troy

터키의 랍세키 마을 지나는 곳의 풍경이다.
A scene of Lapseki Village in Turkey.

트로이 목마

Tory Trojan horse

우리가 너무나 잘 알고 있는 '트로이 목마'는 그리스 군과 트로이 군과의 전쟁에 관한 내용으로 만들어졌다. 그리스군은 거대한 목마를 남기고 철수하는 위장 전술을 폈는데, 여기에 속아 넘어간 트로이군은 목마를 성 안으로 들여 놓고 승리의 기쁨에 취하였다. 새벽이 되어 목마 안에 숨어 있던 오디세우스 등이 빠져 나와 성문을 열어 주었고 그리스군이 쳐들어와 트로이성은 함락되었다. 여기서 비롯된 '트로이의 목마'는 외부에서 들어온 요인에 의하여 내부가 무너지는 것을 가리키는 용어로 쓰이게 되었다.

'Trojan horse' so well known to us was made with the contents of war between Greek army and Trojan Army. Greek soldiers had camouflage tactic withdrawing by leaving a huge wooden horse and Trojan soldiers were exhilarated in victory by bringing in the wooden horse into castle as they were deceived. At the dawn, Odiseus and others who were hiding inside the wooden horse came out of the horse to open the gate of castle and Greek army attacked to destroy the Trojan castle. So the terminology of 'Trojan horse' was used as the expression for the internal destruction by outside force.

AUST RALIA

면적은 769만 2208㎢, 인구는 1988만 명(2003)이다. 인구밀도는 2.6명/㎢(2003)이다. 정식명칭은 오스트레일리아연방이다. 수도는 캔버라이고 공용어로 영어를 사용하고 있다. 북쪽은 티모르해(海), 동쪽은 산호해 · 태즈먼해, 남쪽과 서쪽은 인도양으로 둘러싸여 있다.

The area is 7 million 690 thousand 2208㎢ and population is 19 million 880 thousand people. (In 2003). Population density is 2.6 people/㎢. (In 2003) Official name is Commonwealth of Australia. The capital is Canberra and official language is English. It is surrounded by Timor Sea in the north , by coral ocean and Tasman ocean in the east, and by Indian ocean in the south and the west.

캔버라

기하학적 가로망이 뻗어 있는 전형적인 계획도시이다. 목장 · 농경지로 이용되어 왔으나 1908년 수도로 선정되어 전 세계에서 공모한 도시계획을 바탕으로 1913년에 착공, 1927년에 멜버른에서 수도를 옮겼다.

시드니

대도시권은 서쪽 블루산맥, 북쪽 호크스베리강(江), 남쪽 보터니만(灣)까지 뻗어 있으며, 전국 인구의 약 1/4이 몰려 있는 이 나라 최대의 도시이다.

Canberra

It is a typical planned city in geometric networks of streets. It was originally used as ranches and agricultural farms but in 1908, it was selected as the capital and the city had been started to be built in 1913 based on urban plans submitted from all over the world and in 1927, the capital was relocated to here from Melbourne.

Sydney

The metropolitan Sydney is stretched to Blue mountain range in the west, River Hawksberry in the north, and Botany Bay in the south and is the largest city in the country with the concentration of about 1/4 of total population of the whole country .

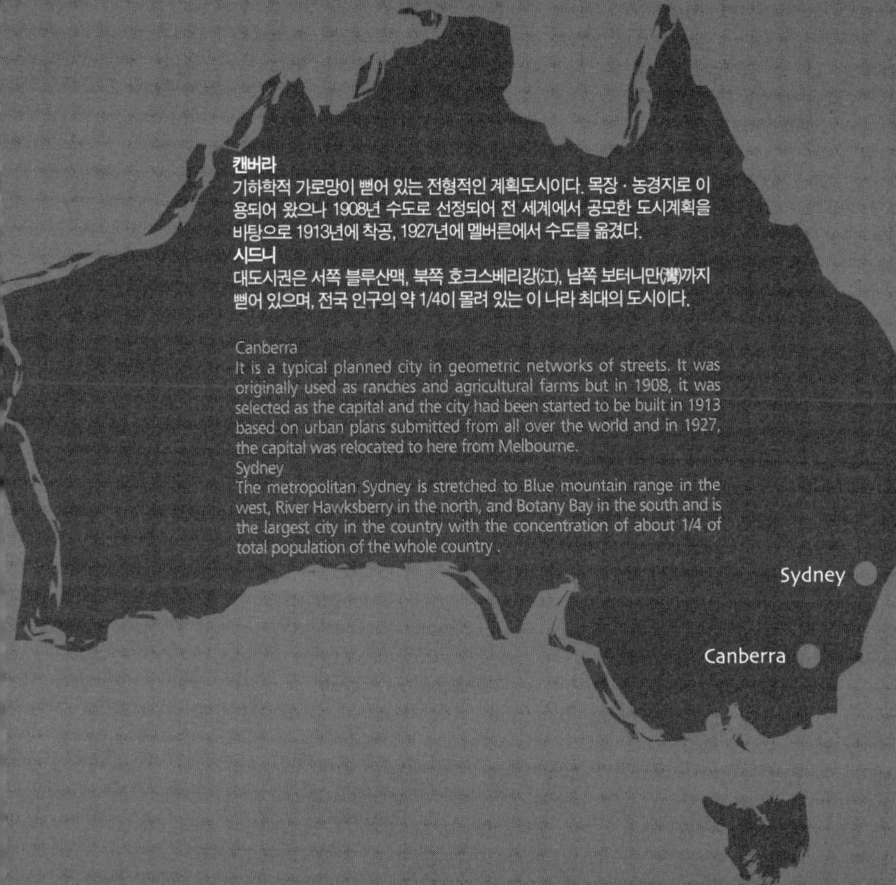

전쟁기념박물관

War Memorial Museum Canberra

1925년부터 계획된 전쟁기념관은 1941년에 완공되어 오늘에 이르고 있다. 기념관에는 제1차 세계대전부터 호주군이 참여했던 모든 전쟁에 관한 사진, 미니어처, 대형폭격기와 전차, 군복, 병사의 일기, 추모사 등 온갖 자료가 방대하게 전시되어 있다. 무려 600만 개가 넘는 조각으로 이루어진 모자이크 벽화와 파란 스테인드글라스가 무척 인상적인 기억의 전당은 제1차 세계대전 때 이름 없이 쓰러져간 군인들의 유골을 1993년에야 찾아내 매장해 놓은 곳이다. 10만 명이 넘는 전몰자의 이름이 새겨져 있어 더욱 수연한 분위기가 흐른다.

War Memorial Museum was planned from 1925 and was completed in 1941. In the Memorial Museum, all kinds of data of all wars Australian army had participated from the World War I such as photography, miniature, large bomber airplane and tank, military clothes, soldiers' diaries, and memorial statements are exhibited broadly. A very impressive Memory Hall with mosaic wall painting made of over 6 million pieces and blue stained glass is a burial place of unknown soldiers of the World War I who were excavated in 1993. Names of more than 100,000 war deaths are engraved to make feel solemn.

오페라 하우스

Sydney The Sydney Opera House

국제공모전에서 1등으로 당선된 덴마크의 건축가 요른 우촌이 설계한 것으로 수많은 지연과 연기, 공사의 어려움으로 점철된 끝에 1973년 완공되었다. 시드니 항구에 정박되어 있는 요트들의 닻 모양을 되살린 조가비모양의 지붕이 바다와 묘한 조화를 이루며, 지금은 시드니를 상징하는 건물로 인식될 뿐 아니라 세계적으로도 유명하다. 하버브리지의 남동쪽에 위치하며, 공연 예술의 중심지로서 극장과 녹음실, 음악당, 전시장을 갖추고 있다. 그리고 시드니심포니오케스트라 및 오스트레일리아 국립오페라단 · 무용단 및 여러 연극단의 본거지이며, 도서관시설도 갖추고 있다.

It was designed by a Denmark architect, Jørn Utzon who won the number 1 prize in international competition for the design and was completed in 1973 after numerous delay, postponement, and difficulties of the construction works. The roof in the shape of shells reviving the anchors of yachts moored at Sydney harbor makes mystical harmony with the ocean and is now recognized as a symbol of Sydney and is also famous world wide. It is located in the southeast of Harbor Bridge and as the center of performance arts, it has theater, recording rooms, music halls, and exhibition halls. And it is the home of Sydney Symphony Orchestra and Australian national Opera and Dance Group and also has library facility.

올림픽 스타디움

Platform in Olympic Stadium Sydney

전체적인 경기장의 기본 설계는 호주 건축 설계 위원회의 주관하에 샌프란시스코의 Hargreves Associates가 하였다. 마스트플랜의 설계개념은 공유지(public domain), 개방공간, 광장, 공원, 도로의 중요성을 주요 건물과 동일하게 다루었다는 점이다. 광장이나 녹지 공간, 도로 등의 공유 공간인 공유지의 설계는 크게 3개의 기능적 구분에 따라 설계되었다.

The overall basic design of stadium was made by Hargreves Associates in San francisco under the auspice of Australian Architecture Drawing Committee. Drawing concept of the master plan had considered the importance of public domain, open space, plaza, park, and road as the same as main buildings. Drawings of public domain such as plaza, green space, and road were designed according to three large functional classifications.

Sydney Gap Park

갭 파크는 더들리 페이지에서 조금 내려간 곳인데, 절벽틈새로 보이는 멋있는 바다경치가 좋다고 하여 갭 파크라는 이름이 붙혀졌다고 한다.

Gap Park is located at a little distance from Dudley Page and it was named as Gap Park since the scenery of ocean between rock cliffs is beautiful.

시드니 하버 브릿지

Sydney Habour Bridge Sydney

시드니 항의 상징이며 싱글아치 다리중에서는 세계에서 두 번째로 긴 다리이다. 해면에서 도로까지의 높이가 약 59m 인 이 대형 다리는 마치 옷걸이 같은 모양을 하고 있어서 '낡은 옷걸이' 라는 애칭을 가지고 있다.
다리의 전체 길이는 1,149m로 뉴욕의 베이욘 다리 보다 약 60cm 짧다. 다리를 지탱하고 있는 파이론 내부를 통해 다리위로 올라 갈 수 있으며 매년 이 곳을 오르는 행사가 사람들에게 즐거움을 주고 있다.

It is a symbol of Sydney harbor and as a single arch bridge, this is the second longest bridge in the world. The height from sea level to road is about 59m and this large bridge looks just like a hanger of clothes and has a pet name of 'old clothes hanger.'
Total length of the bridge is 1,149m and about 60cm shorter than Bayon Bridge in New York. People can go up to the bridge through inside of pipe supporting the bridge and every year, the event of climbing the bridge makes people happy.

opera House
'99.9.26

북쪽은 압록강과 두만강을 건너 중국의 만주와 러시아의 연해주에 접하고, 동쪽과 남쪽은 동해와 남해를 건너 일본에 면하며, 서쪽은 서해를 사이에 두고 중국 본토에 면한다.

The north is faced with Chinese Manchuria and Russian Maritime Province going beyond Abrok river and Duman river and it is faced with Japan in the east and south going beyond the East sea and the South sea. The west is faced with mainland of China between the West sea.

서울

1986년 아시아경기대회, 1988년 서울올림픽경기대회, 2002년 한일월드컵대회가 개최되는 등 국제적인 대도시이다.

Seoul

It is an international metropolitan city that has hosted Asian Game in 1986, Seoul Olympic Game in 1988, and the 2002 Korea-Japan World Cup Soccer Game.

경기도

전국토의 약 10%를 차지하는 크기이다. 해로나 육로로 국도의 남부와 북부를 쉽게 연결해 주며, 한강과 추가령구조곡 등으로 백두대간의 동쪽으로 연결된다. 한강 이북에는 산간지역이, 한강 이남에는 평야지대가 펼쳐져 있다.

Gyeonggi-do

It takes about 10% of the total territory of Korea. Sea route and land transportation connect well with the south and north of national road and is connected to the East of Baekdu Great Range with Han River and Gujo Valley of Chugaryeong. The North of Han river is mountain area and the South of Han river is plain and open field.

인천

한국문화 형성기로 보아 북방과 남방의 중간지대로서 가장 먼저 근대사의 여명을 맞이한 곳으로서 대륙의 문물이 한반도에 전파되고, 한반도의 문물이 해외로 나가는 중요한 관문이었다.

Incheon

As a middle part of the North and the South in the period of formation of civilization of Korea, the dawn of modern ages of Korea had started from this place. This was an important gateway to propagate civilization of the continent into Korean peninsular and to export Korean culture into the world.

K O R E A

충청도

고려 건국 후 995년(성종 14) 전국을 10도(道)로 구획할 때 이 지역을 중원도(中原道)라 하였고, 1106년(예종 1) 중원도에 하남도(河南道)를 병합해서 양광충청도(楊光忠淸道)라 하였다.

강원도

한반도 중앙부 동쪽의 태백산맥을 중심으로 영동과 영서로 크게 구분되어 있다.

전라도

호남은 원래 전라도의 별칭으로 호수의 남쪽에 위치한다는 것에 연유된 명칭으로, 호수는 여러 설이 있으나 김제의 벽골제호를 가리킨다고 전해지며 또한 금강 이남을 강남도라 부른 데에서 호수는 금강을 가리킨다고도 한다.

경상도

1106년(예종 1) 영남 · 영동 · 산남 등의 3도를 합하여 경상진주도(慶尙晉州道)라 하였고, 1171년(명종 1) 이를 양분하여 경상주도(慶尙州道) · 진합주도(晉陝州道)라고 하였다가 1186년에 이를 합하여 경상주도라 하였다. 1204년(신종 7) 상진안동도(尙晉安東道)로 개명하였다가 후에 경상진안도(慶尙晉安道)로 고쳤다.

제주도

면적 1,847.1㎢, 인구 55만 2000명(2002)이다. 북으로 목포와의 거리는 141.6㎞, 북동쪽 부산과의 거리는 286.5㎞이며, 동으로 일본 쓰시마섬[對馬島]과는 255.1㎞ 떨어져 있다.

Chungcheong-do

This place was name as Jungwon (middle) province when the whole nation was divided into 10 provinces during the 14th year of reign of King, Sungjong of Koryeo Dynasty after the foundation of Koryeo in 995 and it was named as Yanggwang Chungcheong Province by merging Hanam Province into Jungwon Province in 1106. (The first year of reign of King, Yejong)

Gangwon-do

Centering on Mountain Taebaek in the East of the center of Korean peninsular, it is divided into two great parts of Youngdong and Youngseo.

Jeolla-do

Honam was a different name of Jeolla and is originated from the place locating in the South of lake. For the lake, there are several theories, it is said that the lake is Byeokgilje lake in Kimje and also is said to indicate Geumgang river as the south of Geumgang river was called as Gangnam Province.

Gyeongsang-do

In 1106, (The first year of reign of King, Yejoong) three provinces of Youngnam, Youngding, and Sannam were merged as Gyeongsangjinjudo and in 1171, (The first year of reign of king, Myeongjong) the province was divided into Gyeongsangjudo and Jinhapjudo and again in 1186, these two provinces were merged as Gyeongsangjudo. In 1204, (The 7th year reign of King, shinjong) it was renamed as Sangjinandongdo and again the name was changed to Gyeongsangjinando later.

Jeju-do

The area is 1,847.1㎢ and population is 550 thousand 2000 people. (In 2002) It is 141.6㎞ away from Mokpo in the north and 286.5㎞ away from Busan in the northeast and 255.1㎞ away from Japanese Tsushima Island (Daema Isalnd) in the east.

서울역

Seoul station Seoul

서울역은 독일 사람인 C.K. 라란데가 설계한 지하1층 지상2층의 르네상스식 건물로 돔과 붉은 벽돌과 흰색 화강석으로 댄 입면은 지금도 그 상징적 가치가 크다고 하겠다. 고속철도, 지하철, 국철 등 폭증하는 교통량으로 인해 지금은 뒷면이 증축되었고 이 역사 건물은 로비공간으로써의 상징적 건물로만 남아있다.

Seoul station was designed by a German architect, C.K. Lalande and is a renaissance style building consisted of one floor of basement and two stories above ground. The dome and the front surface decorated with red color bricks and white color granite still has a high value of symbolism. Now, because of explosive increase of traffic amounts including high speed train, subway, and national railroad, the backside of the station has been expanded and this station building is remained in symbolism as lobby space.

Seoul Novotel Ambassador Fitness Center

휘트니스 센터는 통유리 창 너머로 사계절 변화하는 도심의 경관을 즐기며 쾌적한 마음으로 운동을 즐길 수 있다. 체력장, 최고급 수입산 단풍나무로 바닥재를 설계 시공한 에어로빅 룸, 조깅트랙 등 최첨단 실내 운동 시설과 함께 실내외 골프연습장, 수영장, 남녀 사우나 등을 갖추고 있다. 특히 골프연습장은 강남권의 휘트니스 클럽 중 최고의 시설을 자랑한다. 실내 5타석(5m), 실외 7타석(12m) 규모로, 기호와 날씨에 따라 선택, 이용할 수 있다. 운동 후에는 지압 맛사지 원리를 응용해 근육통과 피로를 해소하고 신체 리듬을 회복해줄 수 있는 썬텐스 스파가 있다.

In this Fitness Center, people can exercise in pleasant mood overlooking the sceneries of mid city that changes in four seasons through wide one piece window. The center is equipped with indoor and outdoor golf practice range, swimming pool, male and female sauna together with the most advance indoor exercise facilities like physical exercise room, aerobic floor designed with the best quality maple wood materials imported, and jogging track. In particular, gold practice range is the best facility among Fitness clubs in the south of Han River. With 5 booths (5m) of indoor and 7 booths (12m) of outdoor practice tees, users can select according to weather and their preference. After exercise, users can use suntan spa where they can relax muscle pains and fatigues and restore body rhythm using the principle of massage.

일진 스포월드

Iljin Spoworld Seoul

스포츠 활동뿐만 아니라 오락과 사교 그리고 비지니스 공간으로서 역할을 수행할 수 있도록 하였으며, 쾌적함과 시원함, 유동적인 물의 이미지와 함께 미래성을 내포하는 디자인에 역점을 두었다. 종합 스포츠센터로써 회원 3,000명을 위한 각 요소별 특징적인 기능을 고려하였으며, 각 시설 이용자들을 위한 원활한 동선 체계를 우선시하였다. 또한 효율적인 평면 및 디테일에 중점을 두어 실용성을 추구했다. 일진 스포월드는 건물 자체가 스포츠레저 전용으로 운영되며, 호텔 수준의 멤버쉽 스포츠센터를 구성하기 위해 최고의 마감재가 사용되어 타 스포츠센터와 고급화라는 차별성을 갖는다.

Iljin Spoworld is designed not only for sport activities but also for entertainment, social activity, and business and the design is focused on comfortableness, freshness, flexible flowing image of water as well as the nature of future. As a comprehensive sport center, characteristic functions per element for 3,000 people of members are taken into consideration and smooth movement line for users was the top priority in design. In addition, practicality was pursued by emphasizing efficient plain and details. Building of Iljin Spoworld itself is operated as exclusive sport and leisure activities and in order to compose the best Membership sport center like the level of Hotel, the best finish materials have been used to be differentiated from other sport centers.

아마랜스 휘트니스 센터

Amaranth Fitness Center Seoul

한남동에 위치한 '아마랜스 휘트니스 클럽' 은 '넓은 주차장, 화려한 실내 인테리어, 연인과 함께, 가족과 함께' 라는 고전적 극장 광고 카피를 떠올리게 하는 곳이다.

여성 전용 휘트니스 클럽인 이곳은 많은 연예인들이 이용하는 곳으로도 유명하다. 헬스는 물론 에어로빅에 골프, 수영까지 모든 것을 즐길 수 있도록 설계된 공간은 놀이동산에 온 듯한 착각이 들 정도다. 그러나 무엇보다 중요한 이 곳의 특징은 남성용 기구가 아닌 여성 전문 휘트니스 기구를 준비해 두고 있다는 것이다. 여성 체형에 맞게 설계되어 훨씬 효과적인 운동을 가능하게 해 주고 있다.

'Amaranth Fitness Club' located in Hannam-Dong reminds a classic advertisement copy of theater of 'ample parking space, luxurious interior design, with lovers, and with family.'

This is exclusive Fitness Club for female and many famous entertainers use the club. The space designed for health as well as aerobic, golf, and swimming makes people to feel they are visiting play ground. But more than anything else, what is important here in this club is that all exercise tools and devices are exclusively for female instead of tools for male. They are designed to fit the physical shape of female and provide more effective result of exercise.

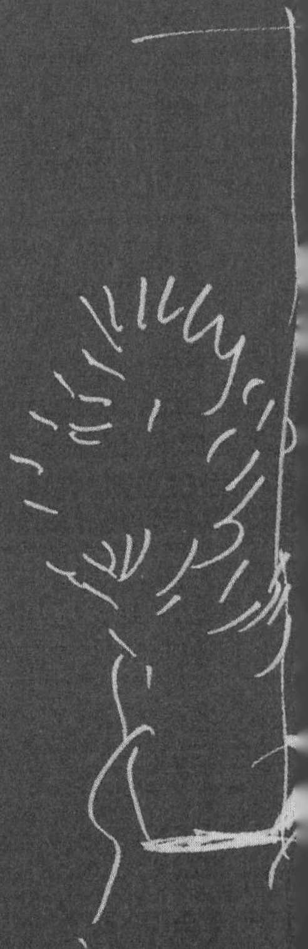

AMARANTH
FITNESS
Center

경희대학교 대강당

Auditorium of Gyeonghee University Seoul

건축 외관의 모티브가 되는 벨기에의 성 미쉘 성당을 재해석하여 대강당 실내 설계에 도입했다. 고딕 양식의 주요 요소인 장미창, 러브앤볼트, 버트레스, 포인티드 아치 등의 디자인 요소를 실내외에서 유기적으로 연결시킴으로, 통일된 이미지를 느낄 수 있도록 했다. 또한 건물의 용도가 다목적 공연홀이라는 기능을 충족시켜야 하므로 고딕 요소의 장식성과 공연장으로서의 기능을 적절히 조화시켜 현대적으로 재해석했다.

They have reinterpreted the motif of St. Michell Cathedral of Belgium to introduce in the interior design of large auditorium. Design elements of rose window frame, love and bolt, buttress, and pointed arch that are all important elements of Gothic style are systematically connected in indoor and outdoor to provide a unified image. In addition, as the usage of the building is for various performances, they have reinterpreted in modern sense by harmonizing properly decorative nature of Gothic style and the functions for performance in order to satisfy the functions.

한국전력 사옥 대강당

Seoul Auditorium of Korea Electric Power Corporation

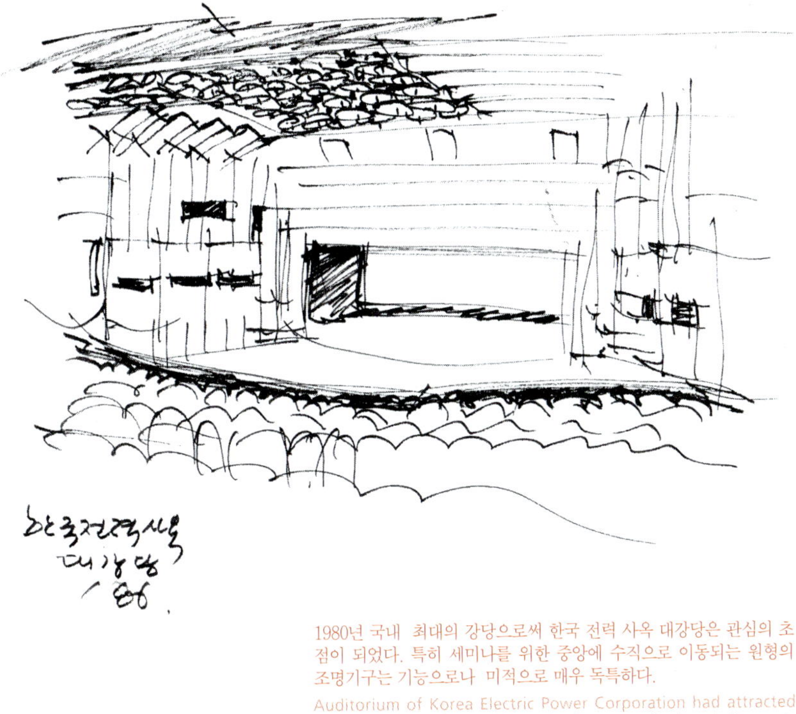

1980년 국내 최대의 강당으로써 한국 전력 사옥 대강당은 관심의 초점이 되었다. 특히 세미나를 위한 중앙에 수직으로 이동되는 원형의 조명기구는 기능으로나 미적으로 매우 독특하다.

Auditorium of Korea Electric Power Corporation had attracted attention of people in 1980 as the largest auditorium in Korea. In particular, a circular illumination device in the center that moves vertically for seminars is very unique in both function and beauty.

동부건설 모델하우스

Dongbu Corporation Model House Seoul

Seoul C.G.V Yongsan

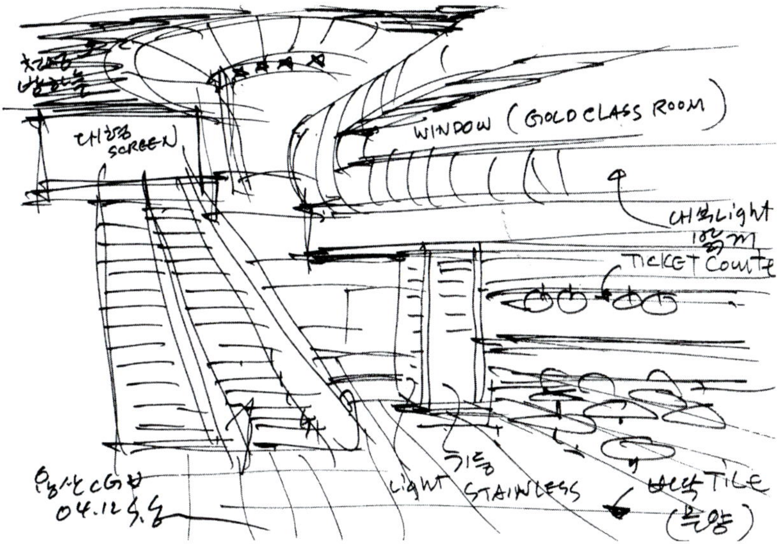

단순한 극장 시설이 아닌 C.G.V만의 독특한 공간연출을 통해 기존극장과 차별화된 느낌을 제공한다. 이를 위해 현실과 다른 현실, 가상세계와 현실을 공유하는 장소, 극장의 네트워크라는 주제로 새로운 개념의 영화관을 제시하였다. 용산역의 이미지를 1백년 후의 미래로 옮겨 우주로 향하는 우주선과 우주공간의 느낌을 표현한 대기홀은 용산이 가지고 있는 장소적 특성을 반영하며, 거대한 알의 형상은 미래 공간을 떠도는 자아를 상징한다. VIP티켓 카운터는 우주여행을 떠날 때 필요한 동력 장치를 형상화했다. 메인 홀의 바닥은 서울을 의미하여, 파란색의 띠는 한강을 의미한다.

Not only as a simple facility of movie theater but through the space representation of only unique to C.G.V, it provides differentiated sense than the existing theaters. For this, it has provided a new concept of networks of movie theaters as the share of a virtual space and reality instead of actual reality. They have moved the image of Yongsan station in the future of 100 years and have expressed spaceship and space of the cosmos in waiting hall to reflect the meaning of place that Yongsan has and the shape of a huge egg symbolizes the self floating in the space of future. VIP ticket counter has symbolized the jet propulsion equipment needed for leaving for space travel. The floor of main hall symbolizes Seoul and blue stripes mean River Han.

C.G.V 불광

C.G.V Bulgwang Seoul

매스를 통한 공간의 볼륨감 조성 - 천정고를 적절하게 조절하였고 동시에 외부 효과를 볼 수 있는 구획을 제공하였으며 높은 천정고를 이용하여 Penthouse 옥외 정원의 높은 천정을 표현하였다. 또한 자연스럽고, 밝고 고급스런 이미지의 마감재를 사용하였다.

Composition of sense of volume in space through mass - the height of ceiling is properly adjusted and simultaneously have provided section that can provide outside effect. Also the high ceiling of outdoor garden on the top of Penthouse is expressed using the ceiling height. In addition, they have used finish materials with the image of bright and luxurious Natural Style.

청룡사 대웅전

Gyeonggi-do Grand Hall, Cheongryong (Blue Dragon) Temple

청룡사 대웅전 Designation : Treasure No. 824

지정사항 : 보물 제824호 Location : 28

소재지 : 경기도 안성시 서운면 청룡리 28번지 Cheongryong-Ri, Seowoon-Myeon, Ansung City, Gyeonggi-Do

1265년(원6) 명본국사가 창건하고 대장암이라 한 후 1364년(공민왕13) 나옹화상이 산이름을 서운산이라 하고 절 이름을 청룡사로 고친 후 오늘에 이르고 있다. 대웅전은 1720년(숙종46) 사간도인이 중수했다고 전하고 있다. 이후 여러 차례의 중수를 거쳐 오늘에 이르고 있다. 1985년에 보물 제824호로 지정되었다. 지붕은 팔작 지붕이며 정면 3칸, 측면 4칸의 평면을 갖고 있다.

It was first built by a patriot, Myeongbon in 1265 (Won 6) as Daejangam and in 1364 (The 13th year of reign of King, Gongmin), Naong had changed the name of mountain to Seowoon mountain and renamed the name of temple as Chyeongryong temple. Main hall is said to have been expanded by a sage, Sagan in 1720. (The 46th year of reign of King, Sukjong) After that, it has been rebuilt several time to have the current feature. It was designated as treasure No. 824 in 1985. The roof is in the shape of Chinese letter of eight and has 3 columns in the front and 4 columns in the side.

안성 정무공 오정방 고택

Old House of Oh, Jeong-Bang, Jeongmugong in Ansung Gyeonggi-do

지정사항 : 유형문화재 제174호, 1650년대
소재지 : 경기도 안성시 양성면 덕봉리 246번지
규모 : 안채 및 사랑채 'ㄱ' 자형 16칸, 문간채 'ㅡ' 자형 4칸, 사당

2000년 3월 24일 경기도유형문화재 제175호로 지정되었다. 원래 덕봉리 252번지에 세워졌었는데 1650년(조선 효종 1)에 현재의 위치로 이전하였다. 1510년(중종 5)에 처음 건립된 이래 정무공 오정방(吳定邦:1552~1625), 천파공 오상(吳祥), 충정공 오두인(吳斗寅) 등 해주오씨의 명현들을 배출한 유서깊은 이 가옥은 안채, 사랑채, 문간채, 사당으로 구성되어 있으며 전체적인 목재 푼수와 치목 상태, 가옥의 배치 및 구성 등이 탁월하다.

Designation : Tangible cultural heritage No. 174, Circa 1650
Location : 246, Deokbong-Ri, Yangsung-Myeon, Ansung City, Gyeonggi-Do
Size : Inner building and Sarang (the male quarter) building 'ㄱ' shape 16 columns, door side building 'ㅡ' shape 4 columns, and ritual temple

It was designated as No. 175 of tangible cultural heritage of Gyeonggi-Do on March 24, 2000. Originally, it was erected at 252, Deokbong-Ri but was relocated to the current place in 1650. (The first year of reign of King, Hyojong, Chosun Dynasty) Since it was first built in 1510 (The fifth year of reign of King, Jungjong), this house had produced several distinguished sages like Oh, Jeong-Bang, Jeongmugong(1552~1625), Oh, Sang, Cheonpagong, and Oh, Du-In, Chungjeonggong. This house has such a profound history of producing famous sages of Oh family of Haeju and consists of inner building, Sarang building, and door side building and overall conditions of wooden materials used and deployment and composition of house are very excellent.

06.6.17

이천 미란다 호텔

Miranda Hotel in Icheon Gyeonggi-do

온천장으로 유명한 이천 미란다호텔이 찜질방, 실내외 수영장, 워터파크 시설을 추가해 수상 테마파크 형태로 새로 개장했다. 기존의 온천과 노천 온천은 시설을 보완해 그대로 살리고 미용탕과 족탕 시설을 추가했다. 히노키탕 · 계곡탕 · 이벤트탕 등으로 구성된 노천 온천은 수영복을 착용해야 하는 곳과 남녀가 분리돼 전라로 이용할 수 있는 곳으로 나뉘어 있다. 노천 온천의 종류도 여럿이고 찜질방 시설이 잘 되어 있어서, 다양한 시설을 취향에 맞게 즐길 수 있다.

Miranda Hotel in Icheon that was famous as a hot spring has reopened as water theme park by adding hot sauna, indoor and outdoor swimming pool, and water park facility. They have maintained the existing hot spring and outdoor spring by complimenting the facilities and had added cosmetic spring and foot spring facility. Outdoor hot springs consisted of Hinoki spring, valley spring, and event spring are divided into public spring where people use in swimming suits and male/female spring where people can use in naked.

As the kinds of hot spring are diverse and with excellent sauna facilities, people can use various facilities according to their preference.

한국가스공사

Gyeonggi-do **KOGAS**

주요업무는 천연가스 제조 · 공급과 천연가스 및 액화석유가스 개발 및 수출입, 천연가스 생산기지와 공급망 건설 · 운영, 관련사업 연구와 기술개발, 부대사업 등이다. 또 천연가스자동차(NGV), 가스냉난방기, 연료전지, NOx저감기술, 냉열이용기술 등을 연구 개발하고 있다.

KOGAS was established in August 1983 and had introduced LNG for the first tie in Korea in October 1986. In February 1987, it had started to supply urban gas to metropolitan capital region and had founded Korea Gas Technology Industry Corporation in May 1993. In November 1995, Korea Gas Maritime Corporation was established and in January 1997, KOLNG was established and was converted as an investment company in October 1997. In March 1998, Korea Gas Engineering was founded and was registered at Korean Stock Market in December 1999 and KORAS was established.

Main businesses are production and supply of natural gas, development and import and export of natural gas and LNG, construction and operation of production facility and supply network of natural gas, research and technology development and related businesses of natural gas. In addition, the company studies and develops natural gas vehicle (NGV), gas heating and cooling devices, fuel cell, NOx reduction technology, and cold heat use technology.

교원공제 인천 회관 레스토랑

KTCU Incheon Restaurant Incheon

Incheon C.G.V Juan

메인 로비는 대지 특성상 좁고 긴 형태로 자칫 답답함을 줄 수 있는 공간이다. 디자인 초기부터 이 점을 고려해 로비 중앙 천장을 최대한 높여 공간을 넓게 보이게 하였다. 화려한 칼라나 조명의 들뜬 느낌보다는 고급스러운 느낌을 주는 조명과 재료로 공간을 계획하였다. 전반적으로 메탈의 굵은 터치로 공간을 심플함과 고급스러움을 강조하였으며, 천장부터 바닥까지 흐르는 벽천은 한결 정갈한 느낌의 공간을 연출하였다. 진입복도는 상영관에 들어가지기 전 고급스러운 유리기둥의 반복으로 화려하지만 동적이기 보다는 마치 대나무 숲을 연상하게 만드는 정적인 분위기를 연출하였다.

Main lobby can be a space to give sense of stifling as it is narrow and long because of the characteristics of land. Considering this point from the initial satge of design, we have secured the ample space by raising the height of ceiling of the central lobby to the maximum. Space was planned with illumination and materials that give rather luxurious mood than bright color or sense of excitement of illumination. In overall, with the thick touch of metals, we have emphasized the simplicity and luxuriousness of space and have represented clean and neat mood from ceiling to floor. The entrance lobby before going into movie room is expressed in splendid but static than dynamic mood of going through bamboo forest with the repetitive luxurious glass pillars.

진천 보탑사 3층 목탑

3 Stories Wooden Tower, Botap temple in Jincheon Chungcheong-do

소재지 : 충북 진천군 진천읍 연곡리 483

보련산 기슭에 위치한 보탑사는 통일을 염원하기 위해 세워진 3층 목탑이다. 보탑사 3층 목탑으로는 우리나라 최대 규모로 상륜부까지 탑의 전체 높이만 42.7m이다. 현재 우리나라에 현존하는 목탑으로는 법주사 팔상전, 쌍봉사의 대웅전이며 세 번째가 이곳 보탑사 3층 목탑이다. 보탑사의 특이한 점은 사방을 돌아가며 각 층마다 다른 이름의 현판이 걸려 있고 사람이 직접 올라 갈수 있다. 대웅보전을 비롯하여 사방불전, 대장경전, 미륵전등 사찰 구조물이 모두 탑 안에 모여져 있다.

Location : 483, Yeongok-Ri, Jincheon-Eub, Jincheon-Gun, Chungbuk

Botap temple located in the foothill of Boryeon mountain is 3 stories wooden tower built praying for the unification. As 3 stories wooden tower, it is the highest in Korea with the total height of 42.7m. This is the third wooden tower existing in Korea nowadays including Palsang Hall of Beobju Temple and Grand Hall of Ssangbong temple. The unique point of Botap temple is that in each floor in all four directions, each different sign board in different name is hung and people can climb up in person. All temple structures including Grand Hall, Sabang Buddha Hall, Buddist Sutras Hall, and stone Buddha are merged within the tower.

Chungcheong-do National museum of Cheongju

소재지 : 충북 청주시 상당구 명암동 87번지

박물관 건물은 건축가 김수근 선생이 설계한 한국 현대건축의 기념비적인 작품으로 우암산 동쪽 기슭에 경사지를 이용하여 조화롭게 배치되어 있다.

상설전시실에는 충청북도에서 출토된 선사시대로부터 조선시대에 이르는 2,300여점의 유물을 시대별로 전시하여 충북지역의 역사와 문화를 살펴볼 수 있다. 야외에는 진천 석장리유적에서 조사된 백제시대의 제철로와 청주 용담동 유적의 통일신라시대 무덤을 복원 · 전시하였다.

Location : 87, Myeongam-Dong, Sangdang-Gu, Cheongju City, Chungbuk

Museum building was designed by an architect, Kim, Soo-Geun as a monumental modern architecture of Korea and is well arranged in harmony by using the slope in the esat of Uam mountain.

In permanent exhibition hall, over 2,300 pieces of ruins excavated in Chungcheongbuk-Do ranging from prehistoric era to the era of Chosun Dynasty are exhibited per period and we can view the history and culture of Chungbuk regions. In outdoor, iron furnace of Baekje Dynasty investigated in the ruins of Sekjang-Ri, Jincheon and tombs of unified Shilla period from the ruins of Yongdam-Dong, Cheongju are restored and exhibited.

보은 법주사

Beobju Temple, Boeun Chungcheong-do

소재지 : 충북 보은군 속리산

이 절의 개조(開祖)로 알려진 의신(義信)이 일찍이 불법을 구하러 천축(天竺: 인도)으로 건너가 그 곳에서 경전(經典)을 얻어 귀국하여 나귀에 싣고 속리산으로 들어가 553년(신라 진흥왕 14) 이 절을 창건하였는데, 법(法)이 안주할 수 있는 탈속(脫俗)의 절이라 하여 법주사라는 명칭이 붙여졌다고 한다. 그러나 법주사의 정신적 지주가 된 미륵신앙(彌勒信仰)이나 법상종(法相宗)의 유식사상(唯識思想)은, 혜공왕 때 이 절의 중흥(重興)에 크게 기여한 진표(眞表)와 그의 제자 영심(永深)에 의하여 발현(發顯)된 것이라 할 수 있다. 성덕왕 때 중수(重修)하였는데, 지금 남아 있는 석물(石物)은 모두 이 때 만들어진 것이라 한다. 그 후 여러 차례 중수를 하여 현존하는 목조건물은 모두 조선 후기의 것이다.

경내에는 유명한 법주사팔상전(法住寺捌相殿: 국보 55)을 비롯하여 쌍사자석등(雙獅子石燈: 국보 5), 석련지(石蓮池: 국보 64), 사천왕석등(四天王石燈: 보물 15), 마애여래의상(磨崖如來倚像: 보물 216) 등의 국보 · 보물을 비롯하여 원통보전(圓通寶殿), 세존사리탑(世尊舍利塔), 사천왕문, 희견보살상(喜見菩薩像), 복천암 수암화상탑(福泉庵秀庵和尙塔), 복천암 학조등곡화상탑(學祖燈谷和尙塔) 등의 지방문화재가 있다.

Location : Sokri mountain at Boeun-Gun, Chungbuk

Euisin who is known to open this temple had visited India early to obtain Buddha law and returned to Mt. Sokri after obtaining the Buddha Sutra and putting on the back of donkey and had founded the temple in 553 (The 14th year of reign of King, Jinheung of Shilla). And the name of Beobju was given as law can stay in safety away from real world of earthliness. However, the religion of stone Buddha that had become the spiritual foundation of Beobju temple or ideology of knowledge of the school of Beobsang religion were all realized by Jinpyo and his pupil, Youngsim who had contributed a lot for the re-building of this temple during King, Hyegong. It was rebuilt during King, Seongdeok and stone items remained now were all made at that time. After that, the temple had been rebuilt several times and currently remaining wooden buildings were all built during later half of Chosun Dynasty.

Within the temple, there are many national treasures and treasures such as famous Palsang Hall of Beobju temple(National treasure 55), twin lion stone lamps (National treasure 5), stone lotus lake (National treasure 64), Four heaven kings stone lamp(Treasure 15), and Maaeyeorae Buddha staue (Treasure 216), and regional cultural heritages such as Wontong treasure hall, Sejon sarira stupa, four heavens king door, Heuigyeon Buddha statue, Suamhwasang tower and Hakjodeunggokhwasang tower of Bokcheonam.

법주사 06.6.18

법주사 팔상전

Palsang Hall, Beobju Temple Chungcheong-do

충청북도 보은군 내속리면(內俗離面) 법주사 경내에 있는 조선시대의 목조 건물.
1962년 12월 20일 국보 제55호로 지정되었다. 우리나라 유일의 목조 5층탑으로, 높이는 22.7m이다. 법주사는 553년(신라 진흥왕 14)에 창건되었고, 팔상전은 정유재란 당시 불에 타 없어진 후 선조 38년(1605년)부터 공사를 시작하여 인조 4년(1626년)에 완성된 것으로, 1968년의 해체 복원 공사를 거쳐 현재의 모습을 하고 있다. 벽의 사방에 각 면 2개씩 모두 8개의 변상도(變相圖)가 그려져 있어 팔상전이란 이름이 붙었다.
전체적인 형태는 정사각형의 돌로 만든 기단부 위에 목조로 5층 탑신부를 쌓고 가장 위쪽에는 철제로 만든 상륜부를 두고 있다. 기단은 돌로 낮게 2단을 쌓았고, 그 위에 갑석을 얹었으며 4면에는 돌계단이 있다. 탑신은 1층과 2층은 정면과 측면이 각 5칸, 3층과 4층은 정면과 측면이 각 3칸, 5층은 정면과 측면이 각 2칸씩으로 되어 있다.

This is a wooden building during Chosun Dynasty located within the temple of Beobju at Naesokri-Myeon, Boeun-Gun, Chungcheongbuk-Do.
It was designated as national treasure No. 55 on December 20, 1962. This is the only 5 stories wooden tower in Korea and its height is 22.7m. Beobju temple was built in 553 (The 14th year of reign of King, Jinheung of Shilla) and Palsang hall was destroyed in fire during Jeongyujaeran and was rebuilt starting in the 38th year of reign of King, Seonjo (In 1605) and was completed in the 4th year of reign of King, Injo (In 1626). It was dismantled and reassembled in 1968 to have the current shape. On each wall of four surfaces, each 2 pictures of different shape are painted and was named as Palsangjeon. (eight shapes)
The overall feature consists of the regular square base made of stone and on top of that, 5 stories of wooden tower were built and on the utmost top is made of steel. The foundation layer was stacked in two layers of stones and covering stone was placed on top of the foundation and has stone steps on all four sides. Tower consisted of each 5 columns on the first and second floor in the front and side, each 3 columns on the third and fourth floor in the front and side, and each 2 columns on the fifth floor in the front and side.

팔상전, (속리산, 법주사 (5층탑))
06.6.18.

보은 선병국 가옥

Old House of Seon, Byeong-Gook, Boeun Chungcheong-do

1984년 1월 10일 중요민속자료 제134호로 지정되었다. 1919~1921년 하개리 마을에 지은 전통가옥으로 선민혁이 소유 · 관리한다. 속리산에서 흘러내리는 삼가천(三街川)이 큰 개울을 이루고 개울 중간에 삼각주를 이루어 섬이 된 연화부수형(蓮花浮水形) 명당이며, 아름드리 소나무들이 숲을 이룬 중앙에 99칸의 큰 기와집이다. 안채 · 사랑채 · 사당의 3공간으로 되어 있으며 안담으로 둘러싸고 다시 바깥담으로 크게 둘러쌌다. 대문은 솟을대문이고 오른쪽과 왼쪽으로 행랑채가 서 있다. 행랑채 끝에는 사랑채로 들어가는 중문채로 이어지고 중문은 솟을삼문형이다. 사랑채는 남향으로 무사석같이 다듬은 세벌대 위에 있다.

It was designated as important folk data No. 134 on January 10, 1984. It is a traditional Korean house built in 1919~1921 at Hagae-Ri village and is owned and managed by Seon,Min-Hyeok. River Samga flowing out of Mt. Sokri makes a big stream and delta is formed in the middle of the stream as an island and is a propitious site of water floating lotus style and it is a huge clay tile roof house with 99 rooms in the middle of tall and huge pine forest. It consists of three spaces of inner building, sarang building, and ritual temple and is enclosed with inner wall and again enclosed with outer walls again. The main gate is a tall gate and there are door side building on both right and left side of the gate. Toward the end of door side building, it is connected to the middle gate leading to sarang building and the middle gate is also a tall triangle gate. Sarang building is facing the south and is located on the three layers of stones formed like stone without sands.

추사고택

Old House of Chusa Chungcheong-do

1976년 1월 8일 충청남도유형문화재 제43호로 지정되었다.
조선 후기의 대표적인 실학자이며 서예가인 추사 김정희의 고택이다. 원래는 99칸집이었다고 하나 지금은 면적 265㎡ 가량이며 안채와 사랑채 · 문간채 · 사당채가 있다. 동쪽에 사랑채를, 서쪽에 안채를 배치하되 안채 대청의 방향은 다른 고택들과는 달리 동향이고 사랑채는 남향이다. 안채는 6칸 대청과 2칸의 안방과 건넌방이 있고 부엌과 안대문 · 협문 · 광 등을 갖춘 ㅁ자형의 집이다.
안방 및 건넌방 밖에는 각각 툇마루가 있고 부엌의 천장은 다락으로 되어 있다. 사랑채는 ㄱ자로 되어 별당채와 같은 성격을 띠고 있다. 지붕은 긴 홑처마에 팔작지붕으로 지형에 따라 기단의 높낮이가 생긴 곳에서는 맞배지붕으로 층을 지게 처리하였다. 사랑채의 함실 부분에도 맞배지붕에 이어붙인 지붕이 기능적으로 처리되어 있다.

It was designated as tangible treasure No. 43 of chungcheongnam-Do on January 8, 1976.
This is an old house of Chusa, Kim, Jeong-Hee who was a representative realist as well as a famous calligrapher of the later part of Chosun Dynasty. Originally, it was a house with 99 rooms but itis now about the size of area of 265㎡ and consisted of inner building, Sarang building, door side building, and ritual temple. In the east, Sarang building is located and in the west, inner building is located but the direction of large floor of inner building is facing the east differently from other old houses and Sarang building faces the south. Inner building consisted of a large floor of 6 columns and two inner room and opposite room and is in shape of ㅁ equipped with kitchen, inner gate, narrow gate, and barn.
Outside of inner room and opposite room, there are each verandah and the ceiling of kitchen is formed as a loft. Sarang building is in the shape of ㄱ and is like a separate building. Roof is formed in different height according to the topography in Chinese letter of eight and at the different points of the roof, crossing roof is added in different layer. At the part of furnace below a room, connected roof is added in functional treatment.

1542년(중종 37) 풍기군수 주세붕(周世鵬)이 고려의 유현(儒賢) 안향(安珦)의 사묘(祠廟)를 세우고 다음해에 학사(學舍)를 이건(移建)하여 백운동서원(白雲洞書院)을 설립한 것이 이 서원의 시초이다. 그 후 1544년 여기에 안축(安軸)과 안보(安輔)를, 1633년(인조 11)에는 주세붕을 추배(追配)하였다. 1550년(명종 5) 이황(李滉)이 풍기군수로 부임해 와서 조정에 상주하여 소수서원이라는 사액(賜額)과 《사서오경(四書五經)》《성리대전(性理大全)》 등의 내사(內賜)를 받게 되어 최초의 사액서원이자 공인된 사학(私學)이 되었다. 1871년(고종 8) 대원군의 서원철폐 때에도 철폐를 면한 47서원 가운데 하나로 지금도 옛 모습을 그대로 간직하고 있다.

The first erection of lecture hall was Baekun-Dong lecture hall established by Ju, Se-Hoon, the head of Punggi County who had relocated tomb and relocated the school building in commemoration of Anhyang, a confucian sage of Koryeo Dynasty in 1542. (The 37th year of reign of King, Jungjong) After that, Anchuk and Anbo were added in 1544 and in 1633 (The 11th year of reign of King, Injo), Ju, Se-Bung was added in commemoration. In 1550 (The 5th year of reign of King, Myeongjong), Lee, Hwang came to be the head of Punggi County and asked royal palace to grant the official sign of Sosu Lecture Hall and also received 《Four Books and Five Classics of Confucianism》 and 《Great Books of Principle of Human Nature》and it had become the first officially granted Lecture Hall and private school. In 1871 (The 8th year of reign of King, Kojong), when Deawongun had abolished lecture halls in the country, 47 lecture halls were saved and it was also one of those saved lecture hall and maintains the original features even now.

日新齋
紹修書院
'98.4.19
學求齋

봉정사

Bongjeong Temple Chungcheong-do

경북 안동시 서후면(西後面) 천등산(天燈山)에 있는 통일신라시대의 절이다.
대한불교조계종 제16교구 본사인 고운사(孤雲寺)의 말사이다. 672년(문무왕 12) 의상(義湘)이 창건하였다. 창건 이후의 뚜렷한 역사는 전하지 않는다.
경내에는 대웅전 · 극락전 · 고금당(古今堂) · 화엄강당(華嚴講堂) · 해회당(海會堂) · 적연당(寂然堂) · 객료(客寮) · 양화루(兩化樓) · 장경고(藏經庫) · 동암(東菴:靈山菴) · 서암(西菴:知照菴) · 덕휘루(德輝樓) 등의 중요 건축물과 고려시대의 대표적 석탑인 3층석탑이 있다. 이 가운데 극락전은 국보 제15호, 대웅전은 보물 제55호로 지정되어 있다. 또한 서암에는 《독포도덕(獨抱道德)》이라는 선조 어필의 현판이 있고, 극락전 · 대웅전에는 《대장경》 판목이 보관되어 있다.

This is a temple during the period of unified Shilla located at Mt. Cheondeung, Sehu-Myeon, Andong City, Gyeonggbuk.
This is a branch temple of the headquarters, Goun temple of the 16th district of Korean Buddhism Jogye Religion. It was erected by Euisang in 672. (The 12th year of reign of King, Munmu) The history after the foundation is not known clearly.
Within the temple, there are many important architectures such as grand hall, paradise hall, old time hall, Hwaeom auditorium, ocean meeting hall, quite hall, guest quarter, Yanghwa tower, Sutras barn, the east cottage, the west cottage, and Deokhwui tower and also there is a representative 3 stories stone tower of Koryeo Dynasty. Among those, paradise hall is designated as national treasure No.15 and grand hall is designated as treasure No.55. In addition, in the west cottage, there is royal calligraphy sign of 《Single Embracing Morals》by King, Sunjo and wooden prints of 《Great Buddhist Scriptures》are kept in paradise hall and grand hall.

鳳停寺 '98.4.19.

쌍데힐 골프클럽하우스 락커룸

Locker room of Golf Club House, Ssangdehill Chungcheong-do

충북 장호원에 위치해 있다.
기존 클럽하우스의 리노베이션에서 신축의 의미로 재해석 되었다. 자연의 아름다움을 세련되고 감각적으로 표현함으로써 클럽하우스의 품격을 한층 높이고자 하였다. 공간의 실용성을 극대화하고 단아하고 품위있는 디자인으로 폭 넓은 연령층에게 공감대를 이끌어 내고자 하였다.

It is located at Janghowon, Chungbuk.
In the renovation of the existing club house, it was re-interpreted as the concept of new construction. Grace and quality of club house was improved by expressing the beauty of nature in refined sensibility. Practicality of space was maximized and it tried to attract attentions of broad level of ages with elegant and graceful design.

Chungcheong-do Spavis in Asan

온천수를 이용하여 건강증진과 휴양을 도모하기 위해 만들어진 스파비스는 크게 바데풀과 실외 온천풀, 남여 사우나, 그리고 건강전문식당 등 각종 부대시설로 이루어져 있다. 바데풀은 한방 클리닉에서 각 개인별 특성에 맞게 수중운동 및 수치료를 할 수 있는 대형 원형풀로 설계되었으며 실외 온천풀은 유수풀, 가족탕, 동굴폭포, 연인탕, 건강보도, 워터슬라이드 등 온가족이 함께 즐길 수 있는 시설로 구성되어 있다. 온천장은 영역에 따라 바닥과 천정고에 차이를 두고 특히 6시 높이의 천정 일부를 천정으로 처리, 자연빛을 실내에 적극적으로 유입하여 개방감을 느낄 수 있도록 하였다.

Spavis made to improve health and resting using water of hot spring consists of largely body pool and outdoor hot spring pool, male/female sauna, and other auxiliary facilities such as health professional restaurant. Body pool is designed in a large circular pool where people can exercise in water and in water treatment according to individual characteristic with oriental medical clinic and outdoor pool consists of facilities that all family can enjoy such as flowing water pool, family pool, cave water fall, lovers pool, health walking path, and water slide. Hot spring has different height in floor and ceiling per section and in particular, we have tried to bring in natural sunlight positively into indoor so that we feel the sense of openness by treating the ceiling at the height of 6 as a part of ceiling.

아산 스파비스

Spavis in Asan Chungcheong-do

Spavis made to improve health and resting using water of hot spring consists of largely body pool and outdoor hot spring pool, male/female sauna, and other auxiliary facilities such as health professional restaurant. Body pool is designed in a large circular pool where people can exercise in water and in water treatment according to individual characteristic with oriental medical clinic and outdoor pool consists of facilities that all family can enjoy such as flowing water pool, family pool, cave water fall, lovers pool, health walking path, and water slide. Hot spring has different height in floor and ceiling per section and in particular, we have tried to bring in natural sunlight positively into indoor so that we feel the sense of openness by treating the ceiling at the height of 6시 as a part of ceiling.

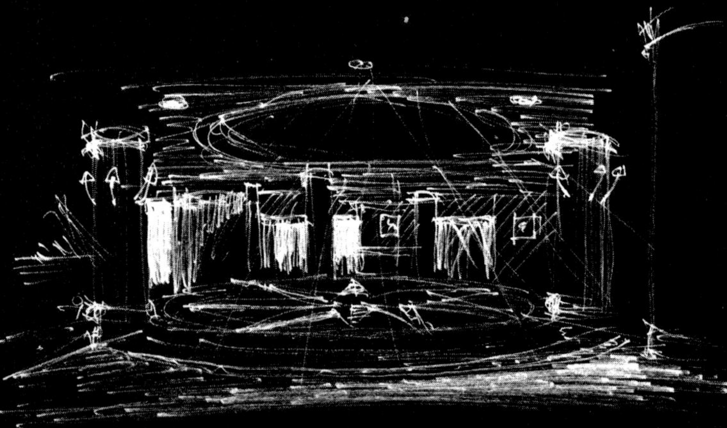

Chungcheong-do Spavis in Asan

Dalma Peak, Mt. Seolak Gangwon-do

"달마" 라는 뜻은 달마대사의 준말로서 달마대사는 선종의 시조이며 남인도 향지국(香至國) 왕의 세째 아들이다. 달마봉은 신흥사 앞의 세심천 건너 동편에 솟아있는 봉우리로 해발 635m의 암봉이며 우리가 흔히 볼 수 있는 달마 대사 그림의 모습처럼 산봉우리의 둥글기가 흡사 달마와 같다하여 달마봉이라는 이름이 붙여졌다.

"Dalma" is an abbriviation of Dalma Great Sage and Dalma great Sage is an ancestor of Sun religion and is the third son of King of Hyangji Nation of the south India. Dalma peak is the peak with the height of 635m above sea level erected in the east across River Sesim in front of Shinheung temple. Just like pictures of Dalma Great Sage we often see, as the shape of roundness of peak of the mountain range looked just like Dalma, it was named as Dalma peak.

객사문

Gangwon-do Guest quarter gate

1962년 12월 20일 국보 제51호로 지정되었다. 규모는 앞면이 3칸, 옆면 2칸으로 비교적 작고, 양식은 단층 맞배지붕인 주심포(柱心包)집이다. 기단(基壇)은 앞면이 조금 높고 정면에 돌계단을 두었으며, 초석은 몇 가지 형태의 돌을 모아서 앉혔다. 기둥은 앞 · 뒷줄이 둥근 배흘림기둥이고 가운뎃줄의 것은 반듯한 네모인데, 여기에 각각 문을 다는 장치가 있다.

It was designated as National Treasure No. 51 on December 20, 1962. The size is relatively small with 3 columns in the front and 2 columns in the side and the style is in symmetrical roof in one story as Center line of Columns house. Base foundation is a little high in the front and stone steps are placed in the middle of the front and cornerstones were made of several different shapes of stones. Pillars were round and upper part slim shape in the front and back rows and the middle pillar was regular square and devices to hang doors was attached to here respectively.

강릉 선교장

Seongyo House in Gangreung Gangwon-do

1967년 4월 18일 중요민속자료 제5호로 지정되었다. 이 지방 명문으로 알려진 이내번(李乃蕃)이 처음으로 살기 시작하여 대대로 후손들이 거처하는 집인데, 열화당(悅話堂) · 안채 · 동별당(東別堂) · 활래정(活來亭) 등 모두 4채가 있다. 가장 오래된 안채 주옥(住屋)은 당초에 주거를 정한 때의 건물이라고 전하나 확실하지 않다.
사랑채인 열화당은 1815년(순조 15)에 오은거사(鰲隱居士) 이후(李后)가 건립한 것이며, 활래정은 그 이듬해 세운 것을 증손인 이근우(李根宇)가 현재의 건물로 중건하였고, 동별당은 약 50년 전에 새로 건립한 건물이다. 안채는 동편에 있으며 주옥의 평면구조는 ㄱ자형으로서, 동쪽 끝이 부엌이고 건넌방은 서쪽에 있다. 안채와 행랑채 사이에는 담을 쌓아서 막았으며, 행랑채는 남쪽에 있고 서쪽으로 사랑채에 출입하는 솟을대문이 있다.

It was designated as an important folk data No. 5 on April 18, 1967. Known as one of powerful families in local community, Lee, Nae-Beon first lived in this house and his offsprings lived in the same house generation by generation. The house consists of 4 quarters of Yeohwa quarter, inner quarter, the east quarter, and quest quarter. The oldest inner quarter of living room is known as one of the original buildings but it is not quite sure.
As a sarang quarter (male quarter), Yeolhwa quarter was constructed by Lee, Hu, a hermit sage in 1815. (The 15th year of reign of King, Sunjo) and Hwalrae quarter was built in the next year by Lee, Geun-Woo, a great grand son and the esat quarter was built newly about 50 years ago. Inner quarter is located in the east and plain structure of main living quarter is in the shape of ㄱ and the end of the esat is kitchen and opposite rooms are in the west. Inner quarter and servants' quarter are separated by walls and servants' quarter is in the south and a high gate was established before entering into sarang quarter in the west.

경포대 해수욕장

Public beach in Gyeongpodae Gangwon-do

시의 중심가에서 북쪽으로 6km, 경포대에서 1km 지점에 있다. 경포호(鏡浦湖)와 바다 사이에 생성되어 있는 모래로 된 해안으로, 6km의 백사장이 펼쳐져 있고 주위에 소나무숲이 우거져 있다. 부근에는 경포대와 오죽헌(烏竹軒) 등 많은 명승고적이 있다.

It is located 6km away in the north from the downtown and 1km away from Gyeongpo tower. This is sand beach between Lake Gyeongpo and the ocean and 6km of sand beach is stretched with pine forest in the vicinity. in the neighborhood, there are many old famous sightseeing objects like Gyeongpo tower and Ojukheon.

강릉 가족호텔

Gangwon-do Family Hotel

하늘과 바다사이 리조트

Sky & Sea Resort Gangwon-do

설악 썬밸리 컨트리클럽 콘도미니엄

Condominium of Sun Valley Country club house, Mt. Seolak Gangwon-do

강원도 고성에 위치해 있으며 콘도미니엄의 기능이 원스톱으로 처리되도록 입구, 로비, 연회장, 레스토랑을 주공간으로 두고 탈의실, 사우나, 실내 골프장 등 부대시설을 또 하나의 구역으로 분리 구성하였다.
로비 바닥의 마블패턴과 양쪽 기둥사이로 펼쳐진 대리석, 무늬목의 어우러짐은 현대적이면서도 우아한 부위기를 연출한다.

It is located at gosung, Gangwon-Do and the composition of condominium is divided into main zone that includes the entrance, lobby, feast room, and restaurant and auxiliary facility zone that includes clothing room, sauna, and indoor golf course in order to process the function as one stop.
Marble pattern on the floor and the harmony of marble stone and floor patterns stretching between two columns on both sides represent elegant and modern mood.

현대 성우리조트

Gangwon-do Hyundai Sungwoo Resort

사계절 물놀이가 가능한 전천후 실내수영장이다. 워터슬라이드 및 수영 , 물놀이로 피로해진 근육을 풀 수 있도록 월풀 시설을 갖추고 있다.

This is all seasons indoor swimming pool. There are water slide, swimming pool, and whirlpool facility where you can relax tired muscles from water play.

롯데 전주 시네마

Lotte Cinema Jeonju Jella-do

8

C.G.V 김해

C.G.V Kimhae Gyeongsang-do

'찬란한 가야 왕국을 만날 수 있는 C.G.V'
C.G.V 김해에 가면 성곽에 둘러싸여 마치 가야 왕국에 들어온 듯한 느낌을 가질 수 있다. 김해는 옛 가야 의 흔적을 찾아 볼 수 있는 유서 깊은 곳으로써 문화유산의 향기를 반영하는 극장을 선보인다.

'C.G.V where you can meet splendid Gaya Kingdom'
If you go to C.G.V Kimhae, it feels like that you are in Gaya Kingdom as it is surrounded by castle walls. Kimhae is a historical place where you can find ruins of old Gaya Kingdom and movie theater reflecting the fragrance of culture is being introduced.

제주 하얏트 호텔 카지노

Jeju-do Hyatt Hotel Casino in Jeju

중문 해수욕장 절벽에 그림처럼 자리잡은 하얏트 리젠시 제주. 1985년 국내 유일의 아트리움 양식으로 설계되었으며 건물이 원통형 구조라 어느 층에서든지 천정위로 하늘을 볼 수 있다. 유리로 된 돔형 천정에서 눈부신 자연광이 쏟아지고 폭포의 물소리가 들린다.

Hyatt Regency Jeju Hotel located on the hill of Jungmun Beach like a picture. It was built as the only atrium style for the first time in Korea in 1985 and the structure of building is in circular column style, you can see the sky from any floor. Bright natural sunlights come through the dome ceiling made by glass and you can hear the sound of water falls.

SPAIN

면적은 50만 6030㎢, 인구는 4260만 명(2003)이다. 인구밀도는 84.2명/㎢(2003)이다. 정식 명칭은 Estado Espa?ol이며, 스페인은 영어명이다. 수도는 마드리드이며 공용어로 에스파냐어를 사용하고 있다. 서쪽으로 포르투갈, 북쪽으로 프랑스에 접하고, 남쪽으로 지브롤터 해협을 사이에 두고 모로코와 마주하며 동쪽으로 지중해, 북쪽으로 비스케이만, 북서쪽으로 대서양에 면한다.

바르셀로나

주변지역은 콩 · 카카오 · 커피 · 목화 · 옥수수 등을 재배하는 농업지대이며, 그 상업중심지를 이룬다. 주요 탄전도 부근에 있으며, 주의 석유산업발전과 함께 인구가 급속히 증가하였다.rdo

코르도바

북쪽의 시에라모레나산맥과 남쪽의 시에라네바다산맥 사이를 차지하는 안달루시아 저지는 개발의 역사도 오래되고, 이슬람교도가 전한 관개 · 농경 기술에도 힘입어 농업지대로서 알려졌으며, 코르도바는 그 농산물 집산지이다.

그라나다

옛 그라나다 왕국의 수도였다. 시에라네바다 산맥의 북쪽과 과달키비르강(江)의 지류 헤닐강과 다르로강이 합류하는 높이 670m 지점에 있다. 기후는 연중 온화하며, 하늘이 청명하다. 주변에는 비옥한 농업지대를 이루어, 곡물류 · 채소류 · 아마(亞麻) 재배와 양잠이 성하고, 포도주 · 올리브유 생산이 많다.

The area is 500 thousand 6030㎢ and population is 42 million 600 thousand people. (In 2003) Population density is 84.2 people/㎢. (In 2003). Official name is Estado Espa?ol and Spain is an English name. The capital is Madrid and official language is Espana. It is faced with Portugal in the west, France in the north, Gibraltar strait in the south in the opposite of Morocco, Mediterranean in the east, Biscay bay in the north, and Atlantic ocean in the northwest.

Barcelona

Surrounding areas are agricultural regions cultivating bean, cacao, coffee, cotton, and corn and becomes the center of commerce and industries. Major mines are in the vicinity and with the growth of petrochemical industry, population has grown rapidly.

"미로,미술관"

Cordoba

Andalusia low land located between Sierra Morena mountain range in the north and Sierra Vevad mountain range in the south had a long development history and is known as agricultural area by the help of irrigation and farming technology brought by Islam and Cordoba is the collection place of agricultural products.

Granada

It was the capital of ancient Granada Kingdom. It is located at altitude of 670m where the north of Sierra Nevada mountain range, branch stream of River Genil of River Guadalquivir, and River Darro are merging. Climate is mild all year and the sky is blue. Fertile agricultural farming area surrounds and production of grain, vegetable, flax, and sericulture are popular and produces lots of wine and olive oil.

마드리드

유럽의 수도 중 가장 높은 곳에 있으며, 연강수량 419mm로 건조하다. 스페인의 정치 · 문화의 중심지일 뿐만 아니라 근래에는 산업도시로서의 중요성도 크며, 도로 · 철도 및 항공의 요충지이기도 하다. 인구상으로는 유럽 제4의 대도시이다.

메리다

1년에 2번 뚜렷한 우기가 나타난다. 커피 · 사탕수수 · 밀 등의 집산지이며, 섬유제품 · 담배 · 식물유(植物油) · 가구 등을 생산한다. 험준한 산지를 넘어 북서의 마라카이보 분지와 남쪽의 랴노스(야노)를 잇는 도로의 요충지이다.

세비야

1248년 페르난도 3세에 의해 에스파냐에 속했고 15세기 말에는 신대륙무역의 기지로서 전성기를 맞아 에스파냐 최대의 성당과 대학 등이 설치되었다. 그러나 17세기에는 대형 선박의 출입이 편리한 카디스에 밀려, 점차 쇠퇴의 길을 걸었다. 그 뒤 세비야는 역사적인 도시가 되었으나, 여전히 보수적인 안달루시아 지방의 거점 도시로서 명맥을 유지해왔다.

톨레도

수도 마드리드 남서쪽 70km 지점에 위치하는 관광도시이다. 타호강(江) 연안에 있으며 역사 · 미술적으로는 마드리드를 능가하기도 한다.

Madrid

Among the capitals in Europe, it is located at the highest place and has dry climate with annual rain falls of 419 mm. It is not only the center of politics and culture but also is an important industrial city in these days and is a place of strategic importance for road, railroad, and airplane. In population, it is the fourth larges city in Europe.

Merida

It has two distinctive rainy periods in a year. It is main production area of coffee, sugarcane, and wheat and also produces textile, tabacco, and vegetable oils, and furniture. It is a strategic location connecting to Maracaibo basin in the northwest going beyond rugged mountain region and Lianos in the south.

Sevilla

It was conquered and included in Espana by Fernando III in 1248 and had prime time at the end of the 15th century as a tading post for new Continent and had built the largest church and university in Spain. However, in the 17th century, the city had lost hegemony gradually to Cadiz that was more convenient for passage of large size ships. After that, Sevilla had become a historic city but it still maintains the status as a hub city in the conservative Andalucia region.

Toledo

It is a tourism city located 70km away in the south of the capital of Madrid. Located alongside River Tajo, in history and arts, the city exceeds Madrid.

SPAIN

미로 미술관

Miro Art Museum Barcelona

스페인의 유명한 화가 미로는 야수파, 입체파, 초현실주의 작가로 꿈과 같고 시같은 많은 작품을 남겼다. 이 미술관은 미로 자신이 설계하여, 1975년 9월에 완공되었으며 살아 있을 때부터 바르셀로나의 젊은 예술가들의 창조적인 실험 장소로 개방하였다. 외형(外形)은 현대적 조형미를 추구하여 직선과 원호(圓弧)를 조화시켰으며, 외벽은 흰색으로 단장하였다.

전시장 안은 천장구조를 원호로 하여 자연광선이 굴절되어 들어오게 하였으므로 밝고 명쾌한 감을 준다. 바르셀로나 시가지가 발아래 펼쳐지는, 몬주익 공원의 산언덕에 자리잡아 경관이 아름답다.

A famous painter of Spain, Miro had left many pieces of art works like dream and poem as a Fauvist, Cubist, and Surrealist painter. This museum was designed by Miro himself and was completed in September 1975 and was opened to many young artists in Barcelona from the period of his living as a creative test place. Outside shape is in harmony of strait line and circles in pursuit of modern plastic beauty and outside walls were painted in white color.

Inside of exhibition hall is bright and pleasant as natural sunlights come inside through the ceiling in circle shape. Barcelona city is overlooked beautifully under the feet from mountain hill of Monjuic Park.

in Barcelona
06.2.11
"미로"미술관에서

미스 반 데 로에 파빌이온

Ludwig Mies van der Rohe Pavilion Barcelona

이 건물은 근대건축의 거장인 미스 반 데 로에가 1929년 바르셀로나 국제전시회에 독일관을 디자인하였다. 전시 후 1930년 해체되었으나, 건축적 가치가 조명된 후 1980년 바르셀로나 도시계획위원회의 결정으로 1983년 건설이 시작되어 1986년 원래의 위치에 그대로 복원되었다.
십자형 기둥만으로 지붕을 떠받쳐 구조에서 개방된 자립벽과 스크린이 공간을 자유롭게 한정하고 있으며 동시성을 느끼게 하는 유리와 철을 이용한 공간의 표현과 수공간 등은 그의 건축철학을 말하는 작품으로 남아 있다.

This building was designed as the German Pavillion by the great modern architect, Mies van der Rohe. at the Barcelona International Expo in 1929. After the Expo in 1930, the building was dismantled but the architectural value was reevaluated in 1980 by the urban planning committee of Barcelona and with the decision of the committee in 1983, it was started to rebuild and was restored at the original site in 1986.
Opened self-erecting wall and screen in the structure of supporting the roof only with columns in cross shape limits the space freely and simultaneously the expression of space using glass and metal remains as a work telling his philosophy of architecture.

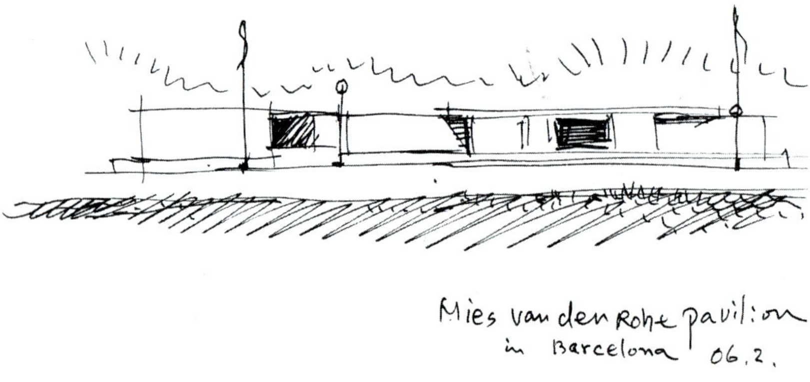

Barcelona Temple de la Sagrada Familia

아르누보의 거장이며 에스파냐의 세계적인 건축가 가우디가 설계하고, 직접 건축 감독을 맡았다. 그의 나이 서른 살 때인 1882년 3월 19일(성 요셉 축일) 공사를 시작해 1926년 6월 죽을 때까지 교회의 일부만 완성하였다. 나머지 부분은 2004년 현재까지도 계속 작업 중에 있으며 교회 전체가 완성되기까지 어느 정도의 시간이 걸릴지는 알 수 없다고 한다. 그는 결혼도 하지 않고 건축에만 매달렸으며, 후에 전차에 치어 숨을 거두었다.

It is a masterpiece in Art Nouveau and designed by Spain's world-renowned architect Antonio Gaudi y Cornet, who directly supervised the architecture himself. The construction began on March 19, 1882 (St. Joseph's Day), when Gaudi was 30 years old, and only a portion of the church was completed until his death in June of 1926. The rest of the church is still under construction today, and it is said to be uncertain when the church will be entirely completed. Gaudi never married and was devoted to architecture, later he was hit by a street car and end his life.

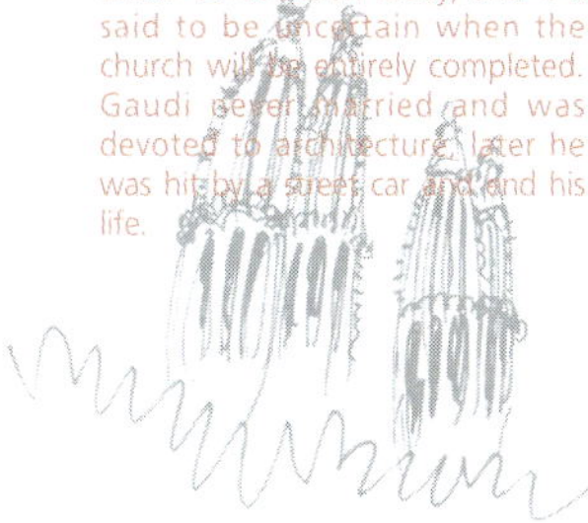

카사바트요

Casa Batllo Barcelona

카사밀라

Barcelona **Casa Mila**

카사밀라는 '라 페드레라' 라고도 하며 1910년 완성하였다. 돌로 지은 5층 건물로서 물결치는 듯 보이는 벽과 복잡하게 얽혀 있는 연철 난간, 동굴처럼 생긴 출입구로 이루어져 있다. 건물 꼭대기는 깬 타일로 만든 모자이크 장식으로 덮여 있고 옥상출입구는 십자가를 토대로 하여 만들었다. 건물 안쪽에는 2개의 안뜰이 있으며 지붕 천창에서 들어오는 빛이 벽면에 투영되어 아름다움을 더하고 있다.

Casa Mila is also known as 'La Pedrera' and was completed in 1910. It is a five story building made out of stone, with undulating walls resembling waves, a complex intertwining wrought iron handrail, and a cave-like entryway. The top of the building is covered with a mosaic of broken tiles, and the entrance to the roof is shaped like a cross. There are two courtyards inside the building, and the light from the opening of the roof ceiling reflects off of the walls, adding more beauty to the building.

몬주익 경기장

Monjuic Stadium Barcelona

몬주익 언덕에 있는 1992년 바로셀로나 올림픽 메인 스타디움으로 우리나라 마라톤의 영웅 황영조 선수가 금메달을 목에 걸었던 곳이다. 중앙에 몬주익 경기장을 상징하는 기념탑은 건축가 칼라트라바에 의해 디자인 되었는데, 마치 미래의 도시, 우주로 날아가는 비행체의 모양으로 방문자들에게 인기가 높다.

As a main stadium of Barcelona Olympic in 1992 located on Monjuic hill, Korean marathoner, Hwang, Young-Jo had held a gold medal on his neck in here. A memorial tower in the center that symbolizes Monjuic Stadium was designed by an architect, Santiago Calatrava and it is very popular to visitors as it is just like in the shape of spaceship flying to the future city, the universe

Barcelona Modern Arts Museum

1988년 4월, 카탈루냐와 바로셀로나시 문화기관, 그리고 현대미술 재단 등이 컨소시엄을 형성해 미술관 건립을 위한 수장품을 모으는 한편 시 당국과 기업의 재정 지원을 얻어 건립 계획을 수립, 미국 태생의 건축가 리차드 마이어에게 설계를 맡겨 5년 만에 개관한 미술관이다. 이 건물은 지하 1층, 지상 3층의 전시동과 7층의 사무동으로 구성되어 있다. 리차드 마이어가 설계한 흰색의 박물관으로 바르셀로나의 오래된 전통 건물과 대비되어 보석처럼 박혀있다. 마이어 특유의 램프를 통한 진입과 백색을 통해 반사된 빛 등이 건물의 특징이다.

In April, 1988, Culture organizations of Catalunya and Barcelona City and Fundacio Museu d' Art Contemporani had formed a consortium to collect storing artifacts for the construction of museum and had established the construction plan with the financial support of city and corporations. The consortium had asked an American architect Richard Meyer for drawings and it was opened after the completion of the construction in five years.

The building consists of one floor of basement, 3 stories of exhibition floors above the ground, and 7 stories of office floors. As a white color museum designed by Richard Meyer, it is like a jewel in contrast to old traditional buildings of Barcelona. The entrance through lamp and reflected lights through white color are characteristics of the building typical to Meyer.

람블라스 거리

Las Ramblas Barcelona

명칭은 아랍어의 람블라 즉, 냇물이라는 뜻에서 붙여졌다고 한다. 람블라스 거리는 바르셀로나의 중심이라 할 수 있는 카탈루냐 광장에서 남쪽의 항구까지 약 1km 남짓 거리를 말하는데, 양쪽으로 차도가 형성되어 있고 가운데에는 이색적인 거리 문화를 볼 수 있는 폭 10m 정도의 도보가 형성되어 있다. 거리 주변에는 호텔, 레스토랑, 오페라극장, 및 시장이 줄지어 있고 플라멩고를 볼 수 있는 작은 소극장바와 피카소 박물관도 그 거리에서 이어진다. 원래 시냇물이 흐르고 있었던 지역으로 약 19세기에 거리로 조성되었다.

The name has an Arabic meaning of Ramblas, that is, a creek. Ramblas street means about the distance of 1km from Cataluna plaza that can be the center of Barcelona to the harbor located in the south and there are pedestrian walk paths with the width of about 10m along vehicle road on both sides for visitors to enjoy exotic street of cultures. In the neighborhood of the street, there are rows of hotel, restaurant, opera house, and market and small theater bars where tourists can enjoy flamingo dance and Picasso museum is also connected through this street. Originally creek was flowing in the region and streets were built in about 19th century.

Barcelona Picasso Museum

람블라스 거리의 중간 위치인 Lucie역 근처에는 피카소 박물관이 있다. 피카소의 유년시절 습작들이 매우 많이 전시되어 있는 이곳은 같은 모델을 정상적으로 표현한 그림과 이를 다시 단계적으로 면과 선으로 분해해가는 과정이 습작으로 전시되어 있다.

There is Picasso museum near Lucie station in the middle of Ramblas street. Trial paintings of youth period of Picasso are exhibited a lot and painting with normal expression and the process of trial paintings of disassembling this normal picture in surface and line in each stage are also exhibited.

해안현대건축

Ocean beach modern buildings Barcelona

람블라스 거리 해안로를 따라 자리잡고 있는 해안 현대 건축물이다.

Coastal modern buildings located along the ocean side road of Ramblas.

Barcelona Housign windows

바르셀로나에 세워진 집합주택의 아름다운 모습이다.

Beautiful windows of housing complexes in Barcelona.

올림피카 빌라

Fish Sculpture Vila Olimpica Barcelona

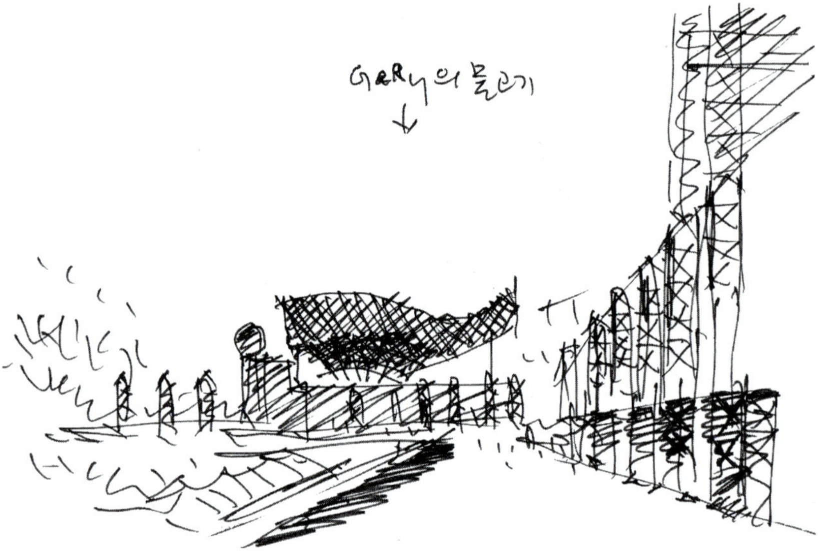

바르셀로나 해변가에 세워진 프랭크 게리의 작품이다. 해체주의 건축가인 그는 규칙과 질서에 얽매여 있는 기존의 사고를 해체시켜 또 다른 형태를 찾고 있다. 이런 해체적 과정을 통해 그는 실험적이고 자유로운 형태를 만들어 나간다.

It is the work of Frank Gehry built in the coast of Barcelona. As an architect with disassemble principle, he looked for another model by dismantling the existing thinking bounded by regulation and order. He had made experimental and free model through these dismantling processes.

파티오

Cordoba Patio

안달루시아 지방의 특징 중의 하나는 집에 파티오(안뜰)가 있다는 것이다. 그리고 파티오를 둘러싸고 방들이 만들어져 있다.

One characteristic of buildings in Andalucia region is that each house has a patio. (inside yard) And rooms are built surrounding the patio.

이슬람 사원

Islam Mosque, Mezquite Cordoba

메스키토는 후기 우마이야 왕조를 세운 아브드 알라흐만 1세가 바그다드의 이슬람 사원에 뒤지지 않는 규모의 사원을 건설할 목적으로 785년 건설하기 시작했다. 붉은색과 흰색으로 띤 대비된 아치는 이곳만의 특징이다. 여러 차례 증축을 거쳐서 지금은 세계 최대의 회교 사원으로 명성을 날리고 있다.

Abuda Al Rahman who had founded later Umaiya Kingdom had started the construction of Mezquite in 785 in order to build the size of mosque as big as Islam mosques in Bagdad. Arch of contrasting red and white color is a unique characteristic of this building. It is now famous as one of the best Islam mosques in the world through several times of renovation and expansion.

메스키타 주변 in 코르도바

알함브라궁전

Alhambra Palace Granada

그라나다를 한눈으로 바라보는 구릉 위에 세운 스페인의 마지막 이슬람왕조인 나스르왕조의 무하마드 1세 알 갈리브가 13세기 후반에 창립하기 시작하여 역대의 증축과 개수를 거쳐 완성되었으며 현재 이 궁전의 대부분은 14세기 때의 것이다. 중정을 중심으로 크게 2개의 아름다운 건물이 배치되어있으며, 분수를 둔 중앙에 둘러싸인 회랑을 따라 아치와, 섬세한 기둥, 이슬람풍의 벽면으로 된 방들이 배치되어 있다.

Muhammed I Al Ahmar. Al Galib of the last Islam Dynasty in Spain, Nasr Kingdom had started to build on a hill overlooking Granada in one glance in later part of the 13th century and was completed through expansion and renovation in every generation and most of currently remaining palaces are from the 14th century. Two huge beautiful buildings are arranged along the central garden and rooms made of arch, delicate pillar, and Islamic style walls are deployed along the corridors surrounding the fountain in the middle.

06.2.15.
- 그라나다 -

알함브라궁전

Alhambra Palace Granada

알함브라 궁전내부
06.2.15, T

저스티스 게이트

Justice Gate Madrid

까를로스 3세가 이태리 건축가 프란시스코 사바티니에게 도시의 기획을 맡겼으며, 그에 의해 도시를 지나는 큰문이 세워졌다. 화강암과 양봉장 근처의 돌을 사용. 나폴레옹 침입 때와 시민전쟁 때의 파편이 아직도 돌에 박혀 있으며 후에는 이 주변을 독립 광장이라 부른다.

King Carlos III had asked urban planning of the city to an Italian architect, Francisco Sabatini and a huge gate that passes the city was built by him. It used granite and stones near beekeeping places. Broken pieces of weapons are still stuck on stones during the attack by Napoleon and civil war and after that, people has called this region as Independence Plaza.

Madrid The Watch Tower(Plaza Mayor)

마드리드의 중심광장인 푸에르타 델 솔에서 5분 정도 걸으면 마요르 광장을 만날 수 있다. 마요르 광장 북쪽 시계탑이 있는 건물 중앙에 'Plaza Mayor' 라는 문장이 새겨져 있고 그 벽에는 세르반테스 등 대표하는 문인들의 초상화가 그려져 있다.

The Watch Plaza Mayor can be reached on foot at about 5 minutes from the central plaza of Puerta del Sol in Madrid. In the middle of the building with watch tower in the north of Plaza Mayor, a crest of 'Plaza Mayor' is engraved and portraits of famous authors such as Cervantes are painted on its wall.

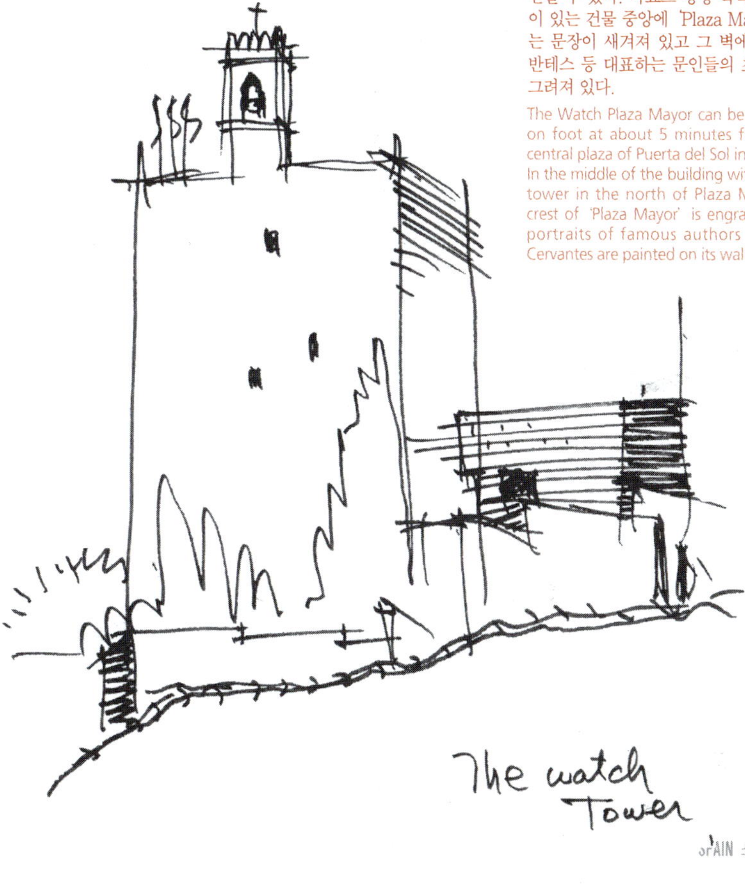

중앙홀

The Main Hall The Museum Merida

Museo Nacional de Art Romano
라파엘 모네오(Estudio Rafael Moneo)건축, 1980~1986

Museo Nacional de Art Romano
Architecture of Estudio Rafael Moneo, 1980~1986

Merida Cathedral of Santamario of la Mayer

Pabellon Mudejar Sevilla

건축가 아니발 곤잘레스가 라틴 아메리카 스타일 스페인 광장에 세운 작품이다.
1893년 산 텔모 궁전의 정원의 반을 도시에 기증한 마리아 루이사 왕비의 이름을 따서 만든 큰 공원에 위치해 있으며, 1929 이베로 아메리카 박람회가 열렸던 곳으로 많은 흥미있는 건물들이 들어서 있는 곳이다.

An architect, Anibal Gonzalez had built Spanish Plaza in a Latin American style .
It is located in a big park built and named after the Queen, Maria Louisa who had donated a half of garden of San Telmo Palace in 1893. It has many interesting buildings as Ibero America Exhibition was held here in 1929.

Sevilla The pavilion Plaza de in Sevilla

스페인 광장

Plaza de Espan Sevilla

건축가 아니발 곤잘레스 작품으로 스페인 광장은 전국에 있는 많은 스페인 광장 중에서도 가장 아름다운 광장 중에 하나이다. 1929년 중남미를 중심으로 했던 이스파노 아메리카 박람회때 본부 건물로 사용됐으며 현재 안달루시아주의 정부 부서들이 입주해 있다. 화장실 타일로 장식된, 무데하르 양식의 스페인 지방의 각 지도와 문장, 그리고 대표적인 역사적 사실들이 타일로 장식되어 있다.

This is the work of an architect, Anibal Gonzalez and it is one of the most beautiful Spanish plazas among the many scattered all over the country. During Ispano American Expo focusing on middle and Latin America in 1929, it was used as Headquarters building and currently, departments of the State of Andalusia government are moved in. It is decorated in cosmetic tiles and each map and crest of Spanish regional communities and typical historical events are decorated in tiles of Mudejar style.

Spain 광장에서
in 세비야
06.2.14

세비아 현대건축

Modern architecture Sevilla

세비야 거리의 현대 건축물이다.
Modern architecture in the streets of Sevilla.

조정학교

Sevilla Rowing school

Rowing School in Sevilla.

세비아 건축대학 사무처

College Office Architecture De Andalucia Sevilla

안달루시아 건축대학 사무처는 중세 14세기에 세비야의 역사적인 중심지였던 부르고스(Burgos)광장 옆에 위치하고 있는 데, 전통적인 세비아의 환경적 제약과 현대적 건축의 요구가 잘 조화되어 있다. 입구에는 4층 높이의 벽만을 세워 외부에서 보면 동의 건물과 높이가 같으나 안으로 들어서면 25%의 공간을 유지하였다. 이 문을 통해 진입하면 본 건물의 현관이 자리 잡고 있고 1층 내부는 수위실과 로비공간이 자리잡고 있으며 지하는 건축서적 중심의 서점으로 꾸며져 있고, 2층부터 6층은 강의실, 제도실 및 연구실로 이루어져 있다.

Headquarters of the College Office Architecture De Andalucia is located next to the historical central plaza of Burgos at Sevilla of the 14th century of the middle ages and traditional environmental limit of Sevilla and modern architectural requirement of buildings are well harmonized. At the entrance, only wall of 4 stories high is erected to match the height of other buildings but if we go into the building, about 25% of space is maintained. When we enter through this gate, main porch of the building is located and inside of the first story, there are doorkeeper office and lobby space and in the basement, there is a book store of architecture. From 2nd floor to 6th floor, there are lecture rooms, drawing rooms, and research rooms.

세비아 수도원,
06. 2. 13

알카사르

Alcazar Sevilla

영화 〈바람과 라이온〉의 무대가 되기도 했던 이 성은 1248년 이슬람교도로부터 세비야를 탈환 한 뒤 14세기에 페드로 1세에 의해 대규모로 개축되었다. 페드로 왕의 궁전 Alcazar del Rey Don Pedro은 여러 개의 방으로 되어 있는데 가장 화려한 곳은 대사의 방 Salon de Embajadores이고, 소녀의 정원 Patil de las Donellas, dlsgud의 정원 Patio de las Munecas을 둘러싼 회랑기둥에 조각된 섬세한 장식무늬가 아름답다.

This castle was the background of the movie of <Wind and Lion> and it was remodeled in a large scale by Pedro I in the 14th century after Sevilla was recaptured from Islam in 1248. As a palace of the King Pedro, Alcazar del Rey Don Pedro consists of many rooms and the most splendid room is Salon de Embajadores. Patil de las Donellas, Dlsgud garden and delicate decoration crest on corridor column surrounding Patio de las Munecas are beautiful.

카테드랄

Catedral Sevilla

이슬람교도를 몰아낸 것을 과시하기 위해 이슬람교 사원 자리에 15세기부터 100여 년에 걸쳐 세운 고딕양식의 대사원이다. 폭 116m, 안길이 76m로, Giralda 탑 옆에 세워진 팔로스 문 Puerta de los Palos로 들어가면 왼쪽에 르네상스 양식의 왕실 예배당 Capilla Real이 있고, 무리요의 그림〈성모수태〉을 소장해둔 회의실 Sala Capitular, 고야와 수브바란 등의 그림을 모아둔 성배실 Sacristiade los Calices이 있다.

It is a great church in Gothic style that was built starting from the 15th century for over 100 years at the location of Islamic mosque in order to demonstrate the power to drive out Islamic believers. When entering Puerta de los Palos with the width of 116m and internal length of 76m built next to Giralda Tower, there is Capilla Real in Renaissance style on the left side and also there are Sala Capitular where the painting of <Conception of Holy Mother> by Murillo is kept and Sacristiade los Calices where paintings of Goya and Zurbaran are kept.

Santa Justa 기차역

Santa Justa station Sevilla

이 세비야역은 스페인이 최첨단을 자랑하는 아토차 (Atocha: '새' 라는 의미) 특급열차의 종착역으로서의 역할을 완벽히 수행하기 위해 역사의 본 건물에서 플랫폼까지 최단 시간에 갈 수 있도록 설계되었다. 플랫폼 천장은 특수 유리로 되어 자연광을 끌어들이고 본 건물 내부로 가능한 한 많은 자연광이 유입되도록 설계되었다.

The station is designed to go in the shortest time from the main building to the platform in order to carry out the role thoroughly as the final destination of cutting edge special train of Spain named Atocha (Meaning of bird) in this Sevilla station. The ceiling of platform is designed and made of special glass to attract natural sunlights as possible as can to the maximum.

세비야
Santajusta 역
06.2.13.

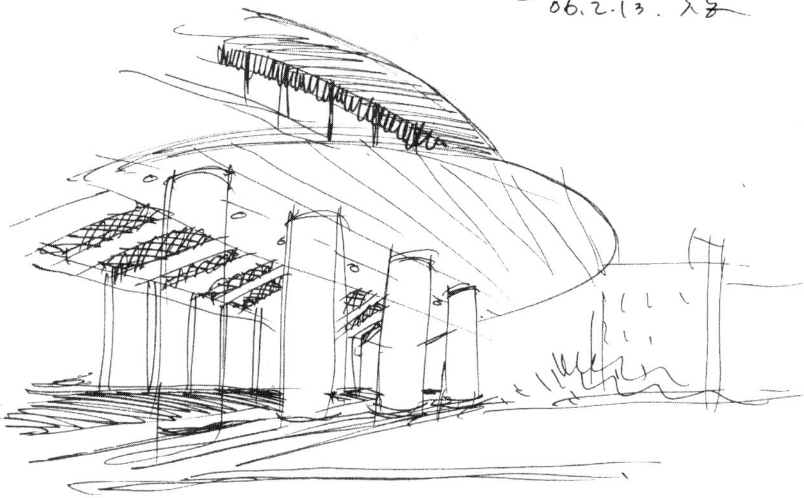

세비야
'06.2.13. Santajusta 역

톨레도 대성당

Toledo Cathedral Toledo

스페인의 이슬람 세력에 승리한 나바스 데 톨로사(Navas de Tolosa)에서 승리한 것을 기념하기 위해 알폰소 8세 때 만들어진 것이다. 높이 92m의 탑에는 18톤의 대형종이 설치어 있으며, 성당 정면에 있는 화려한 조각이 인상적인 세 개의 문을 들어서면 성당 내부로 연결된다. 대 예배당에 있는 제단 병풍은 섬세하고 아름다운 조각으로 유명하다.
13~15세기에 지어진 고딕양식과 아랍양식이 혼합된 성당이며, 세비아의 대성당과 크기가 비슷하며 스페인의 수석 성단으로 사용하고 있다.

It was built during Alponso VIII in order to commemorate the victory in Navas de Tolosa against Islam forces in Spain. A large bell with the weight of 18 tons is installed at the 92m high tower and when entering the three gates with impressive splendid sculptures in the front of Cathedral, it is connected to inside of the Church. A folding screen on the Altar of the great prayer hall is famous for its refined and beautiful sculptures.
It is a cathedral build in the 13th~15th century in mixture of Gothic and Arabic styles and the size is similar to the Cathedral of Sevilla and is now used as prime altar of Spain.

리얼 성당

Toledo San Jeronimuel Real Cathedral

이탈리아반도는 발칸 · 이베리아반도와 함께 남유럽의 3대 반도의 하나로서 동(東)지중해와 서(西)지중해 사이에 있기 때문에 자연 · 문화에 있어서 다른 두 반도와는 달리 순수한 유럽형을 이루고 동시에 지중해 지역의 특색을 가장 잘 나타낸다. 행정구역은 15개주(州)로 되어 있으며 주 밑에 현(縣)을 두고 있다.

ITALIA

Italia peninsula is one of three peninsulas of southern Europe together with Balkans and Iberia peninsula and as it is located between the east Mediterranean and the west Mediterranean, it maintains pure European style unlike the other two peninsulas and simultaneously depicts the best characteristics of the Mediterranean. Administrative districts consist of 15 states and there are prefectures under the states.

과거 고대 로마의 도시이며, 문화의 보고로써 현재는 정치, 관광, 경제, 문화의 세계적 도시로써 유명하다.

The ancient city of Rome is a cultural treasure. It is known for its politics, tourism, economy, and culture, making it a famous global city today.

레오나르도 다 빈치 공항
Leonardo da Vinci Fiumicino Airport

Fiumicino Town Hall
– Italy.

연간 2,000만명을 수용할 수 있는 로마의 중심공항이다. 1961년 개항하였으며 1990년 로마 중심가와 철도를 연결하고 있다. 공항 면적은 1457만 5000㎡로, 활주로는 3,900m×60m 크기 2개와 3,295m×60m 크기 1개가 있으며 수용 능력은 시간당 68회 운항할 수 있다. 계류장은 항공기 77대가 머무를 수 있다. 주차장은 1만 6400대가 동시에 주차할 수 있는 규모이다.
위치는 로마도심에서 서쪽으로 34km에 위치하고 있으며 도심에서 공항으로 가는 교통편은 고속도로 및 전철이 공항까지 직접 연결되어 있다. 남부 유럽의 주요 관문으로서 2005년까지는 신규 여객시설의 건설 및 위성형 터미널 건축, 2층의 6차선 도로 건설 등의 공항 개발 계획을 갖고 있다. 공항은 1974년 설립된 ADR(Aeroporti di Roma SPA:로마공항주식회사)에서 관리 및 운영하고 있다. 운영 시간은 24시간이다.

20 million people pass through this central airport of Rome annually. It opened in 1961 and since 1990 has been connecting central Rome with the railway system(?). The area of the airport is 1457만 5000㎡. There are two runways that are 3,900m×60m, and another that is 3,295m×60m, allowing 68 flights to come in and out of the airport each hour. The airport hanger(?) can hold 77 airplanes. The parking lot can hold 16,400 vehicles at once.
Located 34km south of the center of Rome, and highways and railways provide direct transportation to the airport. As one of the main gateways of souther Europe, new passenger terminals and satellite terminals were constructed in 2005, 2-story, 6-lane road construction, and other plans for airport developments exist. The ADR (Aeroporti de Roma SPA), established in 1974, is in charge of operating and maintaining the airport, 24 hours a day.

면적은 13만 1957㎢, 인구는 1100만 1000명(2003)이다. 인구밀도는 83.4명/㎢(2003)이다. 정식명칭은 헬레니공화국(Hellenic Republic)이다. 반도의 본토와 그 주변에 산재하는 섬들로 이루어져 있으며, 본토의 서안은 이오니아해(海), 동안과 북동안은 에게해에 둘러싸여 있다. 북쪽은 연장 800km에 이르는 긴 국경에 의해서 알바니아, 마케도니아, 불가리아에 접하고, 동쪽은 마리차강(江)을 사이에 두고 터키에 접한다. 발칸반도에 있는 국가로서 수도는 아테네이며, 공용어는 그리스어를 사용한다.

The area is 130 thousand 1957㎢ and population is 11 million 1000 people. (In 2003) Population density is 83.4people/㎢. (In 2003) Official name is Hellenic Republic. It consists of peninsula, mainland and surrounding islands. The west of mainland is surrounded by Ionia ocean and the east and north east coast are surrounded by The Aegean Sea ocean. The north is facing with Albania, Macedonia, and Bulgaria with the extension of 800km long border and the east is faced with Turkey between Maritsa river. As a country in Balkan peninsula, the capital is Athens and official language is Greek.

GREECE

Greece Private House

FRANCE

면적은 54만 3965㎢, 인구는 5977만 3000명(2003)이다. 인구밀도는 109.9명/㎢(2003)이다. 정식명칭은 프랑스 공화국이다. 수도는 파리이며 공용어로 프랑스어를 사용하고 있다. 국토의 모양은 6각형에 가까우며, 3면은 바다로 3면은 산지로 둘러싸여 있다. 동쪽은 이탈리아 · 스위스 · 독일, 북동쪽은 룩셈부르크 · 벨기에와 접하고, 북서쪽은 영국해협을 건너 영국과 마주하며, 서쪽은 대서양, 남쪽은 지중해와 에스파냐로 이어진다.

The area is 540 thousand 3965㎢ and population is 59 million 770 thousand and 3000 people. (In 2003) Population density is 109.9 people/㎢. (In 2003) Official name is French Republic. The capital is Paris and official language is French. The shape of the territory is almost like a hexagon and three sides are surrounded by ocean and three sides are with mountains. It is faced with Italy, Switzerland, Germany in the east, with Luxemburg and Belgium in the northeast, with Great britain in the northwest across the British Strait, with Atlantic ocean in the west, and with Mediterranean and Spain in the south.

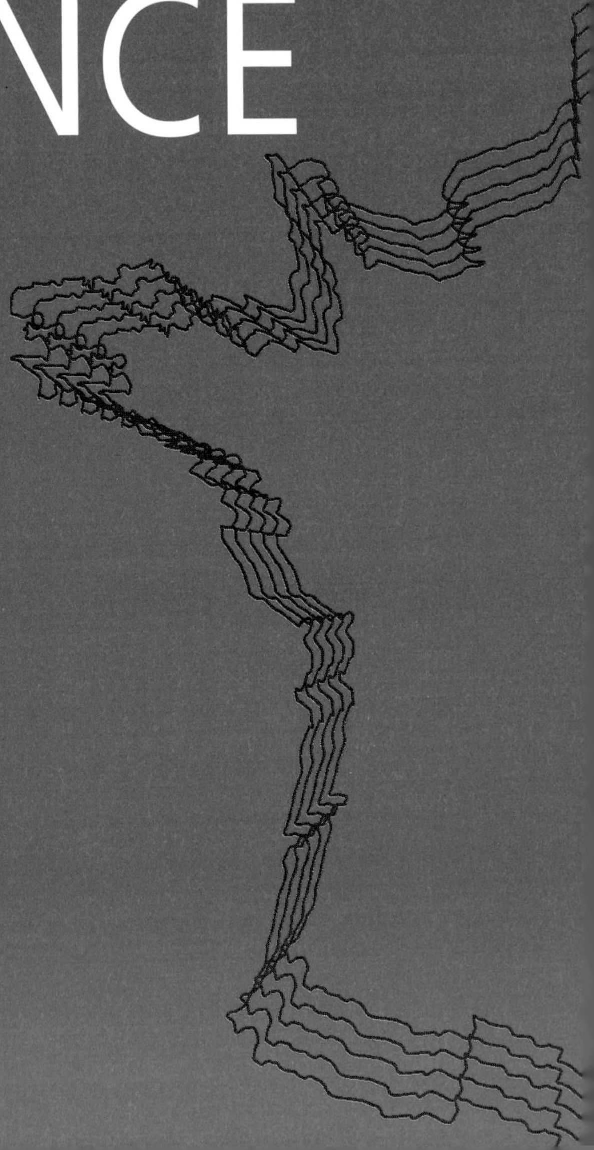

파리

너무나 유명한 문화의 중심지로써 세느 강변을 따라 수많은 관광과 유적지가 산재해 있다.

Paris

At this world famous cultural center, endless sightseeing and landmarks are scattered along the Seine River.

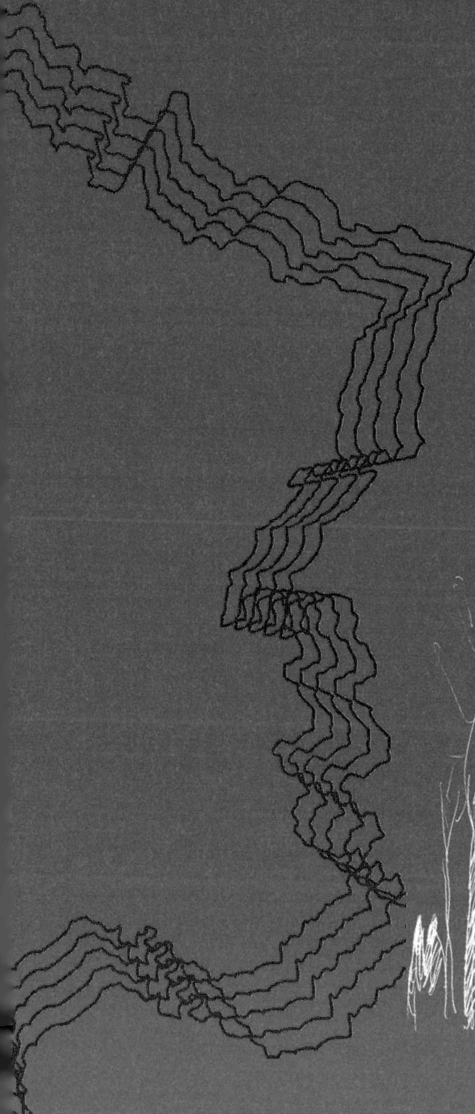

파리공원

Paris Park Paris

세느강

Paris River Seine

길이 776km. 부르고뉴 · 상파뉴 · 일드프랑스(파리 분지) · 노르망디 등을 거쳐 영국해협으로 흘러드는 강으로, 프랑스에서 3번째로 긴 강이다. 세느강을 따라 수많은 도시가 발달해 있는데 특히 파리는 에펠탑, 오르세 미술관, 퐁피두센터, 개선문 등이 강변에서 쉽게 접근할 수 있다.

It is 776km long and is the third longest river in France that flows into British strait via Bourgogne, Champagne, Il de France (Paris basin), and Normande. Numerous cities are developed along River Seine. In particular, in Paris, the Eiffel Tower, the Musee d' Orsay, the Pompidou Center, and the Arc de Triomphe are all easily accessible from the riverside.

노틀담 대성당

Cathedral Notre-Dame Paris

고딕 성당의 대표 건축물이다. 고딕 시대의 화려한 스테인드글라스를 통해 실내로 들어오는 빛은 너무나 화려하고 아름다운데 이를 '장미창' 이라고 한다. 고딕 건축이 하늘로 높아지는 공간을 위해 측면벽에 마치 사람이 서서 벽을 지탱하고 있는 구조물이 있는데 이를 '플라잉 버트레스' 라고 한다. 이 대성당은 나폴레옹의 대관식(1804), 파리 해방을 감사하는 국민예배(1944.8.26) 등 여러 역사적 사건의 무대가 된 곳으로도 유명하다.

This is a typical Gothic cathedral building. Sunlight that comes through splendid stained glasses of Gothic period is so beautiful and it is called as 'rose window.' For the space that Gothic architecture is raised into the sky, there is a structure on side wall surface that just looks like a person is standing to support the wall and it is called as 'flying buttress.' This is famous as the place of holding historically famous events such as the great inauguration of Napoleon (In 1804) and people's prayer for thanking the liberation of Paris. (On August 26, 1944)

Paris Eiffel Tower

파리 국제 박람회를 위해 세워진 이 상징물은 프랑스의 교량기술자 A.G.에펠의 이름을 따서 명명되었다. 높이는 984피트(약 300m)로 파리에서는 어떤 건축물보다 높아 쉽게 찾을 수 있다. 탑에는 3개소에 각각 전망 테라스가 있고 위로 오라가는 엘리베이터를 타고 오를 수 있다. 파리의 경치를 해치는 것이라고 해서 심한 반대가 있었으나 그대로 남아 무전탑(無電塔)으로서 이용되었다. 그러나 제2차 세계대전 후에는 55피트의 텔레비전 안테나가 덧붙여져서 텔레비전의 송신탑으로 사용되고 있다. 탑의 높이는 건설 후 약 40년간 인공 건조물로서는 세계 최고였다.

This symbolic tower built for Paris Expo was named after a French bridge engineer, A.G. Eiffel. The height is 984 feet (About 300m) and can be easily seen in any other buildings in Paris. There are three observatory terraces in the Tower and can go up using elevator. There was a strong protest that the tower hurts the scenery of Paris but is remained as it is and was used as a radio tower. However, after the World War II, it is used as transmission tower of television broadcasting by the attachment of 55 ffet high television antenna. The height was the highest for 40 years in the world after its construction as an artificial building.

Palaris Omnisports de Paris(Bercy Village) Paris

파리시에서 가장 오래된 벡시 마을은 17, 18세기 귀족들의 휴양마을로 알려져 있으며, 또한 프랑스 와인의 집결지였다.

Bercy village that is the oldest in Paris was famous for resort village for royal people during the 17th and 18th century and in addition, it was collection place of all French wines.

Paris EDF Tower

라 데팡스는 파리 중심가 루브르 박물관과 개선문을 중심축으로 도심에서 8㎞ 지점 세느강변에 약 46만평의 면적에 조성된 파리의 부도심이다.
프랑스 정부와 파리시 당국 등 자치단체로 구성된 라 데팡스개발위원회(EPAD)는 지난 58년부터 30여년에 걸친 장기 개발구상을 마련, 80~90년대에 대부분의 공사를 마무리했으며 수많은 관광객이 찾는 파리의 명물이 됐다. 이곳은 도심권에 제약된 건축물 높이가 완화되어 고층, 고밀도 건물이 들어섰으며 이처럼 대단위 지역을 고층, 고밀도로 개발할 때 건물을 비롯해 지하철, 주차장, 일반교통, 보행공간 등을 체계적으로 배치하는 기법을 라 데팡스식 개발이라 한다. 이 지역에 위치한 EDF Tower는 Pei Cobb Freed & Partners에 의해 설계되었다.

La D' efance is the second central city in Paris built in about area of 460,000 Pyeong along River Seine about 8㎞ away from the Louvre museum and Tower of Triumph that are the central area of Paris.
Development Committee of La D' efance (EPAD) consisted of French government and autonomous organizations including Paris city hall had a long-term review period for 30 years from 1958 for the development and had completed most construction works during 1980s~90s and had become a famous place in Paris that so many tourists visit. In here, the height of building that was regulated in the midtown was alleviated and many high-rise and high density buildings were built. And techniques of deploying subway, parking lot, general transportation, and walking space including building systematically when developing the large region like this is called as the development in La D' efnace style. EDF Tower located in this region was designed by Pei Cobb Freed & Partners.

UNITED KINGDOM

면적은 24만 4101㎢, 인구는 5916만 4000명(2003)이다. 인구밀도는 242.4명/㎢(2003)이다. 영국 연합왕국이라고도 한다. 정식명칭은 그레이트브리튼 및 북아일랜드 연합왕국이다. 영국을 흔히 잉글랜드라고 하지만 잉글랜드는 그레이트브리튼의 일부이며, 그 밖에 웨일스와 스코틀랜드가 모여 그레이트브리튼을 형성한다.

The area is 240 thousand 4101㎢ and population is 59 million 160 thoussand 4000 people. (In 2003) Population density is 242.4 people/㎢. (In 2003) It is also called as the British Commonwealth. Official name is United Kingdom of Great Britain and Northern Ireland. United Kingdom is often called as England but England is anly a part of Great Britain and besides, Wales and Scotland form Great Britain.

런던

템즈강을 따라 영국의 정치, 경제, 문화, 교통의 중심지로서 영국의 수도이며 세계적 도시이다.

London

As the center of British politics, economy, culture, and transportation alongside River Thames, London is the capital of United Kingdom and is an international city.

타워 브릿지

Tower Bridge London

런던 앞쪽 템즈 강변에 걸려 있는 런던의 랜드마크이다. 빅토리아풍의 우아한 다리로 1894년에 완성되었다. 다리 위까지는 걸어서 올라갈 수 있으며, 템즈 강변 아래로 펼쳐지는 런던의 풍경을 전망할 수 있다. 특히 선박이 다리 밑으로 통과할 때는 다리가 들어올려지는 모습도 볼 수 있다. 탑 안에는 다리의 구조나 탑의 역사를 전시해 놓은 박물관이 있다.

It is a land mark of London hanging over the Thames in front of London. As an elegant Victorian style bridge, it was completed in 1894. People can walk climbing up to the bridge and can overlook the scenes of London that are deployed under the Thames. In particular, people can see the bridge is lifted up when ships are passing under the bridge. Inside the tower, there is a museum that exhibits the structure or history of the bridge.

버킹검 헌화장소

London Buckingham floral tribute place

버킹검 궁전은 1703년 버킹검 공의 저택으로 건립되었다가 1837년 빅토리아 여왕이 국왕의 궁전으로 바꾸었다. 궁 앞에 있던 거대한 대리석 개선문은 현재는 하이드파크의 북동쪽 모퉁이로 옮겨져 있다. 궁전 앞 빅토리아 여왕의 기념비를 둘러싼 수많은 조각상들은 그 하나하나가 진리, 박애, 정의라는 인간의 기본적인 도덕이념을 의미하며, 과학, 예술, 군사, 공업, 농업의 군상에서부터 기념비 꼭대기에는 승리, 용기, 끈기라는 영국인이 자랑하는 가치관을 상징하는 상이 있다. 이 곳에서 열리는 버킹검의 근위대 교대식은 관광 명물이다.

Buckingham palace was built as the residence for Sir, Buckingham in 1703 and was converted into royal palace in 1837 by Queen Victoria. A huge triumph gate made of marble in the front of the palace is relocated to the northeastern corner of Hyde park. Each of numerous sculptures surrounding the memorial tower of Queen Victoria in the front of the palace signifies basic ideology of human morality of truth, humanity, and justice. The symbols of sculptures range from group features of science, art, military, industry, and agriculture to the moral values of victory, courage, and tenaciousness that british people feel proud at the top of the monument. Shift changing ceremony of the Royal Guards of Buckingham palace is a special sightseeing object.

면적은 4만 1528㎢, 인구는 1623만 8000명(2003)이다. 인구밀도는 479.4명/㎢(2003)이다. 정식 명칭은 네덜란드왕국(Kingdom of the Netherlands)이다. 속령으로 카리브해(海)의 네덜란드령 앤틸리스제도(6개의 섬으로 구성)가 있다. 유럽 북서부에 있는 나라로 수도는 암스테르담이며, 공용어는 네덜란드어이다. 동쪽으로는 독일, 남쪽으로는 벨기에와 국경을 접하며, 서부 및 북부는 북해에 접한다.

NETHERLANDS

The area is 40 thousand 1528㎢ and population is 16 million 230 thousand and 8000 people. (In 2003) Population density is 479.4 people/㎢. (In 2003) Official name is Kingdom of the Netherlands. As subject territory, there are Netherlands Antilles islands (Consist of 6 islands) in the Caribbean ocean. As a country in northwestern Europe, the capital is Amsterdam and official language is Netherlands . It is face with Germany in the east, Belgium in the south, and with North sea in th west and north.

Amsterdam

암스테르담

면적은 637㎢, 인구는 73만 5668명(2002)이다. 네덜란드 서북쪽, 암스텔강이 에이셀호(湖)로 흘러들어가는 지점에 축조된 양항(良港)으로, 북해와는 북해운하로 통하고, 라인강과는 암스테르담-라인 운하로 연결된다. 네덜란드 최대의 무역항으로 13세기에 어민이 암스텔강에 댐을 설치해서 정주한 것이 도시명의 기원이다. 1367년 쾰른 동맹에 참가한 이래, 라인강 무역의 중심지로서 발달했으며, 1602년 자바에 동인도회사를 설립하고 네덜란드가 동양무역을 독점하게 되면서 그 거점으로 번영했다.

Amsterdam

The area is 637㎢ and population is 730 thousand 5668 people. (In 2002) It is a very good harbor built in the merging point of River Amster into Lake Ijsselmeer in the northwest of Netherlands and is connected to North sea through north canal and is connected to River Rhine through Amsterdam-Rhine canal. As the biggest trade port in Netherlands, the name of the city was originated from settlement of fishermen in the 13th century by constructing a dam at River Amster. Since joining Koln Union in 1367, it had developed as central trading place of River Rhine in had established the East India Company at Java in 1602 and as Netherlands became monopolizing the oriental trade, it had prospered as its stronghold.

웰하우스

John Hediuk's Wall House Amsterdam

Amsterdam Expo 2001. Gewild Wonen

면적은 44만 9964㎢, 인구는 895만 8000명(2003)이다. 인구밀도는 21.8명/㎢(2003)이다. 정식명칭은 스웨덴왕국(Kingdom of Sweden)이다. 수도는 스톡홀름이고, 공용어로는 스웨덴어가 사용된다. 스칸디나비아산맥에 있는 긴 국경으로 서쪽은 노르웨이와 접하고, 북동쪽은 토르네강(江)을 경계로 핀란드와 접한다. 동쪽은 보트니아만(灣)과 발트해(海)를 사이에 두고 핀란드 · 독립국가연합과 대하며, 남쪽은 카테가트 해협과 순드 해협을 사이에 두고 덴마크와 접해 있다.

The area is 440 thousand 9964㎢ and population is 8 million 950 thousand and 8000 people. (In 2003) Population density is 21.8 people/㎢. (In 2003) Official name is Kingdom of Sweden. The capital is Stockholm and official language is Swedish. With a very long border in Scandinavian mountain range, it is faced with Norway in the west and with Finland in the northeast bordering with River Torne, with Finland independent nation union in the east bordering between Botnia Bay and Baltic sea Gulf of Bothnia in the east, and with Denmark in the south bordering between Kattegat strait and Sund strait.

SWEDEN

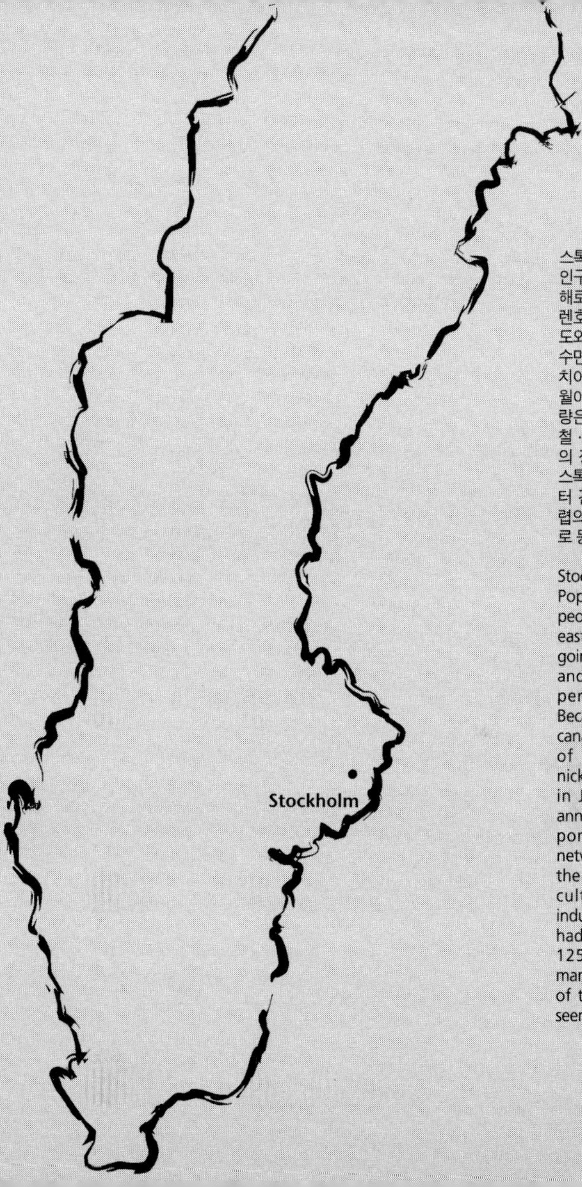

스톡홀름
인구는 75만 8311명(2002)이다. 발트해로부터 약 30km 거슬러 올라온 멜라렌호(湖) 동쪽에 있으며, 시가는 많은 반도와 작은 섬 위에 자리잡고 있다. 넓은 수면과 운하 때문에 흔히 '북구의 베네치아' 라는 별명으로 불린다. 기온은 1월이 −1.6℃, 7월이 16.6℃, 연간 강수량은 555mm이다. 해항 · 공항 · 지하철 · 버스망이 완비되어 있으며 이 나라의 정치 · 문화 · 상공업의 중심지이다. 스톡홀름은 1250년에 스타덴섬에서부터 건설되기 시작했으며, 지금도 그 무렵의 교회와 시장의 광장, 불규칙한 도로 등이 남아 있다.

Stockholm
Population is 750 thousand 8311 people. (In 2002) It is located in the east of Lake Melaren about 30km going upstream from the Baltic sea and the city is situated on many peninsulas and small islands. Because of wide water surface and canal, it is often called as 'Venetia of the North Sphere' in a nickname. Temperature is −1.6℃ in January, 16.6℃ in July, and annual rain falls are 555mm. Maine port, airport, subway, and bus networks are well established and the city is the center of politics, culture, and commerce and industry of this country. Stockholm had started from Stadsholmen in 1250 and even now, church, market square, and irregular street of that time are remained to be seen.

주스웨덴 핀란드 대사관

Finland embassy in Sweden Stockholm

스톡홀름 주스웨덴 핀란드 대사관으로 굴리첸 볼마라라 한다.

As Finland embassy in Stockholm of Sweden, it is called as Gullichsen - Vormara.

시립도서관

Stockholm City library

건축가 아스푸룬드의 작품으로 스웨덴의 민족적 낭만주의와 고전주의적 건물이다. 그의 대표작은 이 건물 뿐 아니라 스칸디아 영화관(1923)등이 있다.

This is the work of an architect, Asplund and is a typical building of Swedish national and classical romanticism. As his representative architecture, there are Skandia movie theater (In 1923) besides this building.

엠일브랙스 교회

Emilbracks Church Stockholm

우드랜드 공동묘지

Stockholm Woodland cemetery (Skogskyrkogarden)

아스푸룬드는 스웨덴에서 당대의 대표적 건축가이다. 그는 레베렌츠를 끌어들여 젊은 나이에도 불구하고 수많은 프로젝트를 담당하게 되었고, '우드랜드 공동묘지' 는 1917년 30대 초반에 공동으로 현상설계에 응모해 당선되었다. 묘지의 건축적 특징과 식물상들을 대지 및 주위 경관의 불규칙성과 결합시킨 디자인으로 하나의 완벽한 기능적 조화를 연출하였다.

Asplund is a representative architect of the time in Sweden. He had participated in numerous projects by inducing Lewerentz despite his young age and he had competed in joint with his colleague designers in public design solicitation for 'Skogskyrkogarden' in 1917 when he was in his early 30s and he had won the competition. He had represented a perfect functional harmony by combining plants and architectural characteristics of the cemetery with the land and surrounding sceneries perfectly as a single design.

귀족의 교회

Burial Chapel Stockholm

시청사

Stockholm City Hall

스톡홀름시의 상징적 건물로 라그나르 오스트베리가 1923년 설계한 이 건물은 20세기의 가장 뛰어난 건물이란 평가를 받고 있다. 내셔널 로만 양식의 건물로 800만개의 벽돌과 1900만개의 모자이크로 지어졌다. 은은하면서도 고상한 기품을 느끼게 해주는 건물에 2층에는 '황금의 방'이 있는데 매년 12월 10일 노벨상 수상자들을 위한 만찬과 파티가 열리는 곳으로 유명하다. 건물의 해변쪽으로는 106m의 높이로 탑이 세워져 탑에 오르면 스톡홀름의 시가지가 한눈에 들어온다.

As a symbolic building of Stockholm, this building designed by Ragnar Ostberg in 1923 is evaluated as one of the most excellent buildings of the 20 century. As a national roman style building, it is built by 8 million pieces of bricks and 19 million mosaics. On the second story of this simple but gracious building, there is a 'golden room' and on the 10th of December of each year, the place is famous for holding dinner and party for Nobel prizes recipients. Toward the ocean side of the building, 106m high tower is erected and people can view the whole city of Stockholm when they climb up the tower.

바로크헤븐 성

Barokhaven Castle Stockholm

노르딕 박물관

Stockholm Nordiska Musset

스톡홀름 시가지의 북부에는 1912년에 만들어진 올림픽 경기장이 있고, 동쪽의 스칸센섬에는 노르딕 박물관과 유원지가 있다. 노르딕 박물관은 르네상스 스타일 성으로 되어있다.

in the north section of Stockholm, there is an Olympic stadium built in 1912 and there are Nordic museum and recreational ground in the eastern Skansen island. Nordic museum (Nordiska Musset) is a renaissance style castle.

면적은 4만 3098㎢, 인구는 538만 7000명(2003)이다. 인구밀도는 125명/㎢(2003)이다. 덴마크어로는 단마르크(Danmark)라고 한다. 정치는 입헌군주제이고 수도는 코펜하겐이며, 공용어는 덴마크어이다. 북해(北海)와 발트해를 가르는 곳에 위치하여, 남쪽으로는 독일과 국경을 접하고, 북쪽으로 스카게라크 해협, 동쪽으로 카테가트 해협을 끼고 스칸디나비아 반도의 노르웨이 · 스웨덴과 대한다. 해외 영토로 페로제도와 그린란드가 있다.

The area is 40 thousand 3098㎢ and population is 5 million 380 thousand and 7000 people. (In 2003) Population density is 125people/㎢. (In 2003) In Danish, it is called as Danmark. Political system is constitutional monarchial system and the capital is Copenhagen and official language is Danish. It is located at the dividing place of the North sea and Baltic sea and is bordered with Germany in the south, with Skagerrak strait in the north, with Kattegat strait in the east and is faced with Norway and Sweden of Scandinavian peninsula. As overseas territory, they have Perrot archipelagoes and Greenland.

DENMARK

코펜하겐

인어공주로 유명한 코펜하겐은 인구는 49만 5699명(2000)이다(수도권은 137만 9천:1998). 덴마크어로는 쾨벤하운이라고 한다. 셸란섬의 북동안에 있는 무역항으로 대안에 있는 스웨덴의 말뫼 사이에는 철도연락선이 오간다. 코펜하겐의 기원은 오래며, 1043년에는 하운 또는 하프니아(항구)라는 이름으로 기록되었다.

Copenhagen

Copenhagen famous for princess mermaid has population of 490 thousand 5699. (In 2000) (The metropolitan region has 1 million and 370 thousand in 1998). In Danish, it is called as København. As a trading port located int he northeastern coast of Sjalland Island, railroad ship is connected to the opposite port of Malmo of Sweden. The origination of Copenhagen is very old and in 1043, it was recorded in the names of Havn or Hafnia. (port)

오페라하우스

Opera House Copenhagen

2004년10월 1일 완공돼 2005년 1월 15일 개관한 '코펜하겐 오페라 하우스'는 덴마크의 선박왕 메르스크 묄러(91). 웁살라 콘서트홀과 덴마크 왕립 극장 리노베이션 설계를 한 덴마크 굴지의 건축가 헤닝 라르센이 건축 설계를, 런던 로열 오페라 하우스 리노베이션을 맡은 영국 애럽사가 음향 설계를 하였다. 운하의 주변 공간에 위치한 이 건물의 유리창을 통해 보이는 코펜하겐의 경관은 너무나 아름답다.

'Copenhagen Opera House' that was completed on October 1, 2004 and was opened on January 15, 2005 was designed in architecture drawings by Maersk Møller(91), a Danish king of ships and the world famous Danish architect, Henning Larsen who had designed Uppsala concert hall and renovation of Danish Royal Theater and also in sound drawings, it was designed by British Arup that had designed the sound system of renovation of London Royal Opera House. Scenery of Copenhagen seen through the windows of this building located in surrounding space of the canal is so beautiful.

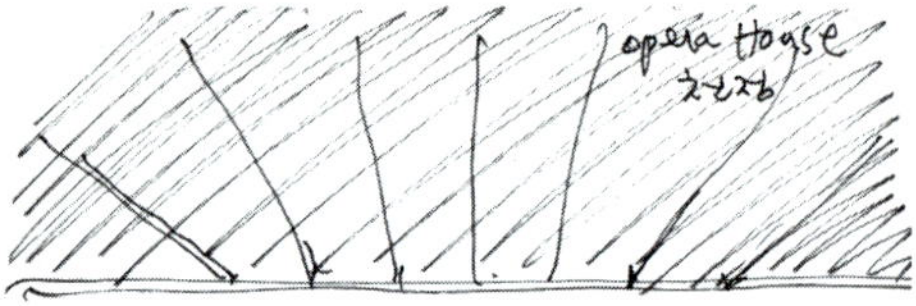

05.9.30 opera House 內部
in copenhagen

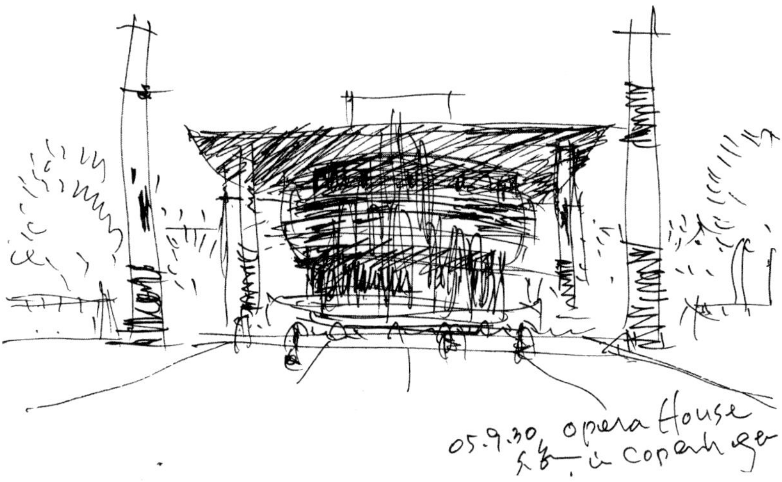
05.9.30 opera House
in copenhagen

프레데릭스버그 성

Frederiksborg Slot Copenhagen

덴마크의 '베르사이유 궁전'이란 별명이 붙은 이 건물은 1602년부터 1630년까지 국왕 크리스티안 4세가 지은 르네상스 형식의 이 성은 '200여 년 동안 7명의 국왕이 이 성에서 대관식을 올릴 정도로 덴마크를 대표하는 성이다. 1859년 화재로 성의 대부분이 소실되었을 때 왕실에서 이를 재건할 경제적 여유가 없자, 맥주 재벌인 칼스버그 야콥센의 기부로 지금의 모습을 갖추게 되었다. 현재는 덴마크의 유구한 역사를 알 수 있는 각종 자료와 회화, 가구, 보물 등을 전시하는 역사박물관으로 사용되고 있다.

This building with a nickname of 'Versailles Palace' of Denmark was built by King Christian IV from 1602 to 1630 in Renaissance style. Over 200 years, 7 kings of Denmark had coronation ceremony in this building and is the most representative palace in Denmark. In 1859, when most of the palace was destroyed by a fire, royal family could not afford the finance to rebuild this palace and by the donation of beer plutocrat, Carlsberg Jacobsen, it was possible to restore as the current feature. Now, it is used as a history museum that exhibit various data, painting, furniture, treasure telling s long history of Denmark.

Castle
in copenhagen
05.8.29

블랙 다이아몬드

The Black Diamond Copenhagen

왕립도서관 옆에 새로 지어진 도서관이다. 검은 돌로 마감되어 블랙 다이아몬드라 불리며 내부는 시원스럽게 통유리로 처리되어 있으며 에스컬레이터로 오르내리는 동선에 서서 보는 운하의 모습이 유리밖에 펼쳐진 한 폭의 그림과 같다.

A new library built next to the royal national library. It is called as the Black Diamond as it is finished with black stones and its interior is treated with whole piece of glass and sceneries of canal seen through glass from the escalator' s movement line are like a piece of painting.

Copenhagen Grundtvig Church

표현주의 건축의 상징물로써 이 교회는1940년 덴마크의 건축가 클린트의 작품으로 외관의 모양은 파이프 오르간을 연상시키며 교회 건물을 상징하고 있다. 아버지인 젠슨 클린트가 죽은 후 그의 아들 카클린트에 의해 완성되었다.

As a symbol of expressionist architecture, this church is the work of a Danish architect Clint built in 1940, the exterior shape reminds pipe organ and symbolizes a church building. The building was completed by son, Care Klint after father Jensen Klint died.

면적은 33만 8145㎢, 인구는 521만 2000명(2003)이다. 인구밀도는 17.1명/㎢(2003)이다. 정식명칭은 핀란드 공화국(Republic of Finland)이다. 핀란드어로는 수오멘 타사발타(Suomen Tasavalta), 또는 수오미(Suomi: 호수의 나라)라고 한다. 보트니아만(灣)과 핀란드만에 면하고, 옛 소련과 스웨덴 사이에 위치한다. 주로 동쪽으로부터 이주해온 피노우그리아계(系)의 언어를 사용하는 핀인(人)과, 나중에 남부로 이주해온 스웨덴인(人)으로 구성되어 있고, 19세기 초까지는 스웨덴의 일부였다.

FINLAND

The area is 330 thousand 8145㎢ and population is 5 million 210 thousand and 2000 people. (In 2003). Population density is 17.1 people/㎢. (In 2003) Official name is Republic of Finland. In Finnish, it is called as Suomen Tasavalta or Suomi. (country of lake) It is faced with the Gulf of Bothnia strait and Finland strait and is located between the old Russia and Sweden. The country mainly consists of Finnish who use the language of finno-Ugrian after relocating from the east and Swedish who had relocated from the south and it was a part of Sweden until the early of the 19th century.

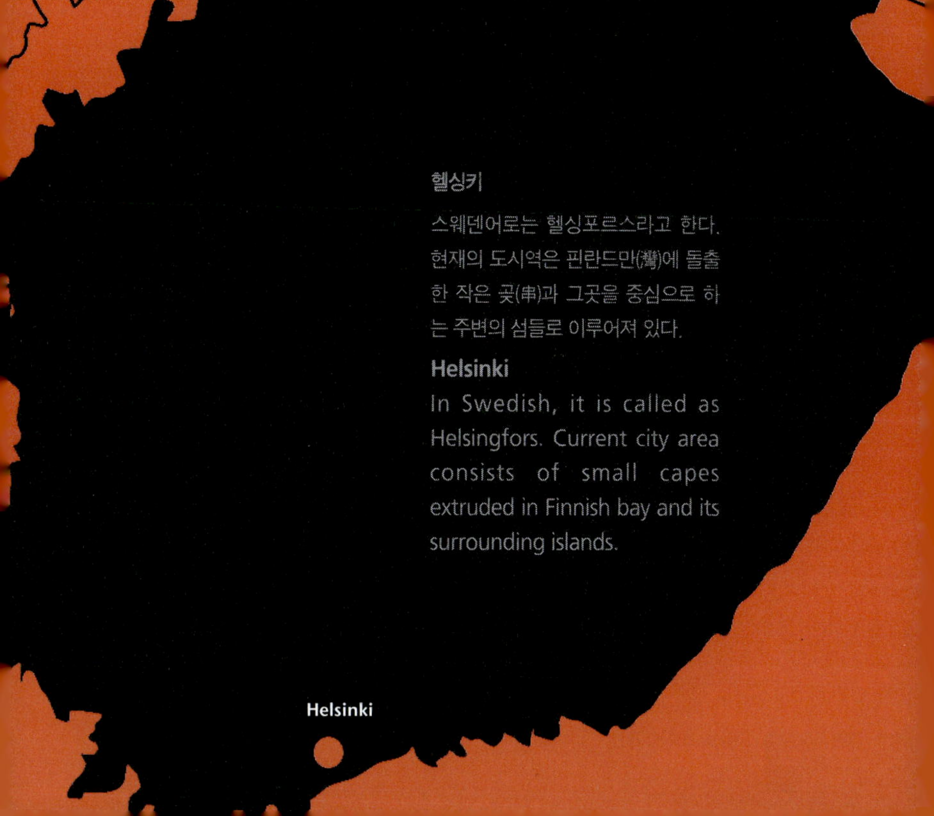

헬싱키

스웨덴어로는 헬싱포르스라고 한다. 현재의 도시역은 핀란드만(灣)에 돌출한 작은 곶(串)과 그곳을 중심으로 하는 주변의 섬들로 이루어져 있다.

Helsinki

In Swedish, it is called as Helsingfors. Current city area consists of small capes extruded in Finnish bay and its surrounding islands.

샤리넨 하우스

Saarinen House Helsinki

19세기 젊은 3인의 건축가(헤르만 게셀리우스, 아르마스 린드그렌, 엘리엘 샤리넨) 저택 중 박물관으로 쓰이고 있는 주 건물은 샤리넨의 저택이다.

Among residence of three young architects (Herman Geselius, Armas Lindgren, Eliel Saarinen) of the 19th century, the main building used as a museum was the residence of Saarinen.

템펠리아우키오 교회

Helsinki Temppeliaukion Kirkko

1969년 수오마라이넨 형제의 설계로, 국립박물관 서쪽 옆의 바위산을 파고 지은 교회이다. 핀란드의 탁월한 건축 디자인을 잘 나타내고 있는 이 교회는, 특히 자연의 음향 효과를 충분히 계산에 넣음으로써 음악회장으로도 자주 활용되고 있다. 3,100개의 파이프를 가진 파이프 오르간에서 퍼지는 음악은 방문객을 압도 한다.
이 교회는 도시의 미관을 해치는 바위언덕을 깨고 핀란드 건축 기술의 진수를 보여주기 위해 만들었다고 한다.

With the design of Suomalainen brother in 1969, the church was built by excavating rock mountain in the west next to the national museum. The church showing the excellency of Finnish architectural design is often used as music performance as, in particular, the effects of natural sound was fully calculated to be included in the design. Spreading music coming out of the pipe organ with 3,100 pipes overwhelms visitors.
The church was built on order to show the essence of Finnish architectural technology by breaking rock mountain that hurts the scenery of the city.

헬싱키 공과대학

Helsinki University of Technology (HUT) Helsinki

헬싱키의 위성도시인 에스포의 오타니에미 라는 곳에 핀란드의 과학기술 메카인 헬싱키 공과대학이다.

Helsinki University of Technology is the Mecca of Finnish science technology located at Otaniemi of Espo that is a satellite city of Helsinki.

콜럼버스 쇼핑센터

Helsinki Columbus Shipping Center

실자라인 선상

Siljaline liner ship Helsinki

스톡홀름과 헬싱키 간을 이동하는데 있어 제일 많이 사용되고 있는 것이 페리로, 이 페리를 타고 바라다보는 핀란드의 풍경은 자연과 오래된 고건축이 조화되어 마치 한편의 영화를 보는 것 같다.

This is a ferry most used for moving between Stockholm and Helsinki and the scenery of Finland riding on this ferry is just like watching a movie as the nature and old architecture are so well harmonized.

Helsinki Helsinki central station (Rautatieaseme)

엘리엘 샤리넨의 설계로 1914년에 완성된 중앙역은 붉은 화강암의 외관과 녹청색 지붕이 인상적이며 역 좌우에는 거대한 조각상이 웅장함을 더하고 있다.
헬싱키 여행의 구심점이 되는 중앙역은 1914년에 지어진 것으로 붉은 화강암을 사용한 외관이 퍽이나 낭만적인 분위기를 풍기고 있다. 이 역의 지하 통로에는 슈퍼마켓과 간이매점, 카페 등이 밀집되어 있다. 중앙역 건너편에는 핀란드 최대의 미술관인 아테네우민 미술관이 있다. 이곳에는 주로 18세기 이후의 미술품들이 전시 되고 있다.

The central station completed in 1914 by the design of Eliel Saarinen has impressive exterior with red granite and greenish blue roof and huge sculptures on both sides of the station add the grandeur of the building.
The central station that becomes the focal point of the travel in Helsinki was built in 1914 and the exterior made of red granite provides very romantic mood. In the underground corridor of this station, supermarket, concession, and cafe are concentrated for the convenience. In the opposite side of the central station, there is the best Finnish art museum, Ateneumin museum. In here, arts after the 18th century are mainly exhibited.

인구는 161만 8468명(1999)이다. 바르샤바평야의 중심부인 비슬라강(江)의 중류 양안에 있다. 13세기에 건설되고, 14세기부터는 그 지방을 차지한 여러 제후령(諸侯領)의 중심도시가 되어오다가 1596년부터 폴란드왕국의 수도가 되었다.

Population is 1 million 610 thousand and 8468 people. (In 1999) It is located between the both sides of middle stream of River Wisla (in the central part of Warsaw). It was built in the 13th century and from the 14th century, it had become the central city of various local leaders who had occupied the region. From 1596, it had become the capital of Polish Kingdom.

● Warszawa

POLAND

면적은 31만 2685㎢, 인구는 3862만 3000명(2003)이다. 인구밀도는 123.5명/㎢(2003)이다. 정식명칭은 폴란드공화국이다. 수도는 바르샤바이며 폴란드어를 공용어로 사용하고 있다. 북동쪽으로 러시아연방, 동쪽으로 리투아니아, 벨로루시, 우크라이나, 남쪽으로 슬로바키아, 체코, 서쪽으로 독일에 접하며, 북쪽은 발트해에 면한다. 남북길이 649km, 동서길이 689km로 거의 4각형의 국토를 이룬다.

The area is 310 thousand 2685㎢ and population is 38 million 620 thousand and 3000 people. (In 2003) Population density is 123.5 people/㎢. (In 2003) Official name is Polish Republic. The capital is Warszawa and people use Polish as official language. It is faced with Russian federation in the northeast, with Lithuania, Belolusi, Ukraina in the east, with Slovakia and Czech in the south, with Germany in the west, and with Baltic sea in the north. The territory is in the shape of regular square with the length of 649km in north and south and 689km in the length of east and west.

빌라노프 궁전

Warszawa Palac Wilanowski

프랑스 베르사이유 궁전을 모델로 하여 지은 이 궁전은 바르샤바에서 남쪽으로 7㎞ 정도 떨어진 빌라노프 지역에 위치한다. 1683년 9월 오스트리아 수도 빈을 오스만투르크군(軍)의 포위에서 구출해 유럽 문명을 구한 군주라는 명성을 얻은 얀 3세(얀 소비에스키)가 부르봉 왕가 출신의 왕비를 위해 여름 별장(궁전)으로 지은 것이다.
17세기에 바로크 양식으로 지어졌으나, 여러 차례 증축되면서 다양한 양식이 가미되었다. 이 밖에 이 궁전의 정원이 명소로써 유명한데 이탈리아 · 프랑스 · 중국식으로 구분해 정원을 배치하였으며, 또 궁전 주위에 넓고 울창한 숲, 주변 환경과 어우러지도록 조성한 인공호수가 있어 건축 · 자연 · 인간이 잘 조화된다. 제2차 세계대전 당시 독일군이 사무실 겸 병원으로 이용했기 때문에 바르샤바의 건물 유적들 가운데 유일하게 전쟁의 해를 전혀 입지 않은 건축물이기도 하다.

This palace that was built imitating French Versailles Palace is located at Wilanow region the south about 7㎞ away from Warsaw. King jan III (Jan Sobieski) who had obtained the fame of saving European civilization by liberating the capital Vienna of Austria from the encirclement of Osman Turk army in September 1683 had built for his queen from royal Bourbon family as her summer resort (Palace).
It was built in Baroque style in the 17th century but had been added various other styles through several times of expansion and renovation construction. Besides, as a famous place of this palace, the garden is very famous and the garden was arranged in division of Italian, French, and Chinese style and also as there are thick forest and man-made lake to match with surrounding environment so the combination of architecture, the nature, and human make a great harmony. Also since the building was used as office and hospital by German army during the World War II, it is the only building among old Warsaw architectures that was not received any damage of the war.

바르샤바 거리

Warszawa Street Warszawa

장례식장

Warszawa Deceased Funeral Chapel

HUNGARY

면적은 9만 3030㎢, 인구는 1013만 6000명(2003)이다. 인구밀도는 109명/㎢(2003)이다. 정식 명칭은 헝가리 공화국이다. 헝가리어로는 머저르 쾨즈타르서샤그(Magyar K)이다. 수도는 부다페스트이고 공용어로는 헝가리어를 사용한다. 북쪽으로 슬로바키아, 북동쪽으로 우크라이나, 동쪽으로 루마니아, 남쪽으로 유고슬라비아, 크로아티아, 서쪽으로 오스트리아, 슬로베니아에 접한다.

The area is 90 thousand 3030㎢ and population is 10 million 130 thousand and 6000 people. (In 2003) Population density is 109 people/㎢. (In 2003) Official name is Republic of Hungary. In Hungarian, it is called as Magyar Köztársaság. The capital is Budapest and official language is Hungarian. It is faced with Slovakia in the north, with Ukraina in the northeast, with Rumania in the east, with Yugoslavia and Croatia in the south, and with Austria and Slovenia in the west.

인구는 182만 1552명(2000)이다. 헝가리 평야의 북서부, 도나우강 양안(兩岸)에 걸쳐 있으며, 우안의 부다와 좌안의 페스트로 이루어져 있다. 부다는 대지(臺地) 위에 자리하며, 왕궁(王宮)의 언덕 · 겔레르트 언덕 등이 강기슭 근처까지 뻗어 있고, 역사적인 건축물이 많다.

Population is 1 million 820 thousand and 1552 people. (In 2000) It is located in between the northwest of Hungarian plain and the both sides of River Donau and consists of Buda on the right shore and Pest on the left shore. Buda is situated on plateau and hill of palace and Gellert hill are stretched to the near site of foothill of river and there are many historical architectures.

국회의사당

Capitol Budapest

다뉴브 강변의 네오고딕 양식의 궁전 같은 건물이 바로 국회의사당이다. 건국 1000년을 기념하여 1884~1902년에 건축되었다. 뾰족 지붕이 하늘을 찌를 듯이 서 있으며 건물 벽에는 헝가리 역대 통치자 88명의 동상이 세워져 있다. 건물 내부에는 691개의 방이 있고 앞의 광장에는 헝가리의 영웅인 코슈트(Kossuth)와 라코지(Rakoczi)의 동상이 세워져 있다.

Neo Gothic style building at the side of River Danube is the Capitol. It was built in 1884~1902 in commemoration of the 1000th anniversary of the national foundation. A sharp roof is erected like piercing the sky and statues of 88 rulers of Hungarian history are erected on the walls of the building. Inside of the building, there are 691 rooms and in the front plaza, the statues of Hungarian heros of Kossuth and Rakoczi are erected.

Budapest Plaza of heros

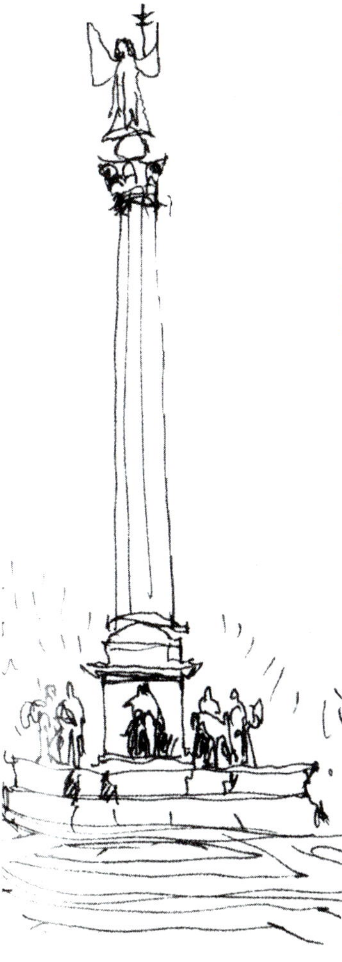

1896년 헝가리 건국 1000년을 기념하여 만들어진 광장이다. 중앙에는 민족의 수호신 가브리엘 천사 동상이 말탄 기사들에 의해 에워싸여 있고 그 주변에는 헝가리의 왕, 정치가, 장군 등 유명인의 동상이 주위에 나란히 서 있다. 국가적인 행사장으로 자주 이용되며 행사동안에 기념비는 군인들에 의해 호위된다.

It was built in commemoration of the 1000th anniversary of Hungary in 1896. In the middle of the plaza, the statue of angle, Gabriel, the protector of Hungarian race is surrounded by cavalries of knights and in the vicinity, statues of famous Hungarian king, politician, and generals are standing side by side. It is often used for national events and during ceremonies, the monument is surrounded and protected by soldiers.

어부의 요새

Fishermen' s fort (Halaszbastya) Budapest

어부의 요새는 네오로마네스크 양식으로, 뾰족한 고깔 모양의 일곱 개의 타워로 설계되어 있고 각 타워들은 수천년 전에 나라를 세운 일곱 개의 마자르 족을 상징하고 있다. 하얀색의 화려한 성벽과 마차시 교회까지 뻗어있는 계단에 서면 저 멀리 다뉴브 강의 아름다운 풍경이 한눈에 들어온다. 이 요새 이름의 유래에 대해서는 두 가지 설이 있다. 하나는 옛날에 어시장이 있었던 곳이기 때문이라는 설이고, 다른 한 가지는 18세기에 어부들이 성벽에서 적군을 방어했다는 설이다.

Halaszbastya is in neo Baroque style and is designed in 7 sharp towers in the shape of peaked hood and each tower symbolizes 7 Marzar races who had founded country several thousands years ago. When standing on splendid castle walls in white color and stairs stretched to Machasi Church, in a far away, you can get a view of the whole scenery of beautiful River Danube. There are two schools of theories for the origination of this fort. One is that it was fish market in the past and another theory is that fishermen had defended the enemy at the castle walls in the 18th century.

쎄체니 온천장

Budapest Szecheny hot spring

온천의 기원은 약 2000년전, 고대 로마인이 로마식 공중목욕탕(카라카라 욕장)을 지을 때로 거슬러 올라간다. 16세기에서 17세기에 걸쳐 이 땅을 지배했던 오스만 터키 제국의 터키인이 더욱 온천을 발전시켰다고 한다.
1876년에 처음 발견된 이 온천은 세계에는 많은 유명한 온천들이 많지만 헝가리에서도 유명한 온천왕국중의 하나이다. 헝가리는 미네랄이 풍부한 광천수가 솟아나오는 곳이 전국에 135개소나 되고 온천마다 각기 다르고 뛰어난 효능을 가지고 있다.

Origination of hot spring traces back to 2000 B.C. when ancient Roman had built Roman style public baths. (Caracara bath) It is said that Osman Turkey Empire who had ruled this land from the 16th century to the 17th century had further developed hot springs.
This hot spring first found in 1876 had made Hungary as one of the world famous hot spring countries in the world even though there are many famous hot springs in the world. There are 135 places of hot springs with plenty of minerals in Hungary and each hot spring has different and excellent efficacy.

CZECH

면적은 7만 8866㎢, 인구는 1020만 2000명(2003)이다. 인구밀도는 129.4명/㎢(2003)이다.

정식명칭은 체코공화국이다. 수도는 프라하이고 체코어를 공용어로 사용하고 있다. 1993년 1월 1일 체코슬로바키아가 체코와 슬로바키아 2개 공화국으로 분리, 독립함에 따라 탄생한 내륙국이다. 북쪽으로 독일 · 폴란드, 서쪽으로 독일, 남쪽으로 오스트리아, 동쪽으로 슬로바키아와 접한다.

The area is 70 thousand 8866㎢ and population is 10 million 200 thousand and 2000 people. (In 2003) Population density is 129.4 people/㎢. (In 2003) Official name is The Czech Republic. The capital is Prague and people use Czech as official language. As of January 1, 1993, Czechoslovakia was separated and became independent into the two republics of Czech and Slovakia and Czech is a new country located in inland. It is faced with Germany and Poland in the north, with Germany in the west, with Austria in the south, and with Slovakia in the east.

면적은 496㎢, 인구는 118만 6855명(2000)이다. 영어 · 프랑스어로는 프라그, 독일어로는 프라크라고 한다. 체코 중서부, 블타바강(江: 몰다우강) 연변, 라베강(江: 엘베강)과의 합류점 가까운 곳에 있다. 체코 최대의 경제 · 정치 · 문화의 중심도시이다.

The area is 496㎢ and population is 1 million 180 thousand and 6855 people. (In 2000) In English and French, it is called as Prague and in German, it is called as Prag.

It is located in the midwest of Czech, near River Vltava (River Moldau), and at the merging point of River Lave (River Elbe). This is the largest center of economy, politics, and culture of Czech.

찰스 브릿지

Charles Bridge Prague

1342년에 홍수로 부서진 기존의 다리를 대신하여 세워진 이 다리는 찰스4세가 건립했다 하여 찰스 브릿지라 부른다. 파리의 몽마르뜨를 연상케 하며 프라하 관광명소 1호로 꼽히는 이 다리는 다리 전체가 조각품이고 이곳에서 내려다 보이는 블타바강과 고성의 아름다움은 관광객을 압도한다.

This bridge that was rebuilt by Charles IV replacing the old bridge because of the destruction of the bridge by flood in 1342 was named after Charles IV.

It reminds Montmartre of Paris and is regarded as number 1 tourist spot in Prague. The whole bridge itself is a sculpture and the beautiful sceneries of River Vltava and old castles overwhelm tourists.

찰스브릿지에서

Prague Prague Plaza

구시가지 광장은 프라하 성을 제외하고 역사적으로 가장 중요한 곳으로 11세기 때부터 발전되어 상업의 메카로써 시장도 이곳에 열렸다. 또한 이 곳은 1621년 합스부르크가에 대항한 27명의 프로테스탄트 추종 귀족을 처형하였고, 1918년 5월 민족 해방을 위한 투쟁 등 많은 역사적 애환이 서린 곳이기도 하다. 이 곳에는 로마네스크 양식에서 아르누보 양식까지 한 자리에서 볼 수 있는 중세 역사의 향기를 느낄 수 있는 건축물들이 많이 있다.

Old city plaza is the most important historical place except Prague palace and it was developed as the Mecca of commerce from the 11th century and markets were opened in here. In addition, in 1621, 27 high class people who followed protestant against habsburg Haus were hung in this plaza and in May 1918, people fought in the plaza for the independence of the nation and therefore, lots of sorrows and pains of history are buried here. There are many architectures that make people feel about the historical fragrance of middle ages as people can see various styles of architectures from Romanesque style to Art Nuveau style in one place.

면적은 8만 3871㎢, 인구는 805만 4000명(2003)이다. 인구밀도는 96명/㎢(2003)이다. 정식 명칭은 오스트리아 공화국(Republic of Austria)이다. 수도는 빈이고 공용어로 독일어를 사용하고 있다. 빈 북쪽은 체코, 동쪽은 헝가리 · 슬로바키아, 남쪽은 슬로베니아와 이탈리아, 서쪽은 스위스와 독일에 접한다.

비엔나

면적은 415㎢, 인구는 156만 2676명(2001)이다. 영어로는 비엔나(Vienna), 체코어로는 비덴(Viden), 헝가리어로는 베치(Becs)라고도 한다. 도나우강(江) 상류 우안에 있는 유럽의 고도(古都)로, 지금도 중부 유럽에서 경제 · 문화 · 교통의 중심지를 이룬다. 수백 년 동안 대제국의 수도였으며 지리적 이점 때문에 정치의 중심지를 이루었다. 중부 유럽을 동서 방향으로 달리면서 민족 · 기후 · 식생 등의 경계를 이루는 알프스 · 카르파티아 산맥을 관류하고 있는 도나우 강변에 자리잡고 있다.

The area is 83,871㎢ and population is 8 million and 54, 000 people. (In 2003) Population density is 96 people/㎢. (In 2003) Official name is the Republic of Austria. The capital is Vienna and official language is German. The north is faced with Czech, the east with Hungary and Slovakia, the south with Slovania and Italy, the west with Swiss and Germany. The area is 415㎢ and population is 1 million 56 thousand and 2,676 people. (In 2001). In English, it is called as Vienna, in Czech as Viden, and in Hungarian as Becs. As an old city of Europe located at the right side of upper stream of Donau River, the city is still the center of economy, culture, and transportation in the middle of Europe. It was the capital of great empire for several hundreds years and was the center of politics because of the geographical advantages. It is located near River Donau that runs through the Alps and Carphatian mountain range that make border of race, climate, and vegetation by running across the middle Europe in the direction of east and west.

르네상스 호텔

Renessance Hotel Vienna

Vienna Schonbrunn

'쉔부른' 이란 뜻은 '아름다운(쉔: schoen) 샘(부른: brunn')이란 뜻으로, 1619년 마티아스 황제가 사냥 도중 '아름다운 샘 Schonenner brunnen'을 발견한데서 유래한 것이다. 그러나 그 기원은 1569년에 막시 말리안 2세가 수렵장으로 쓰기 위해 동물원과 정원을 만든 데서 기원한다.

'Schonbrunn' mean 'beautiful (schoen) fountain (brunn)' and is originated from the discovery of 'Schonenner brunnen' by King, Matias during hunting in 1619. However, it is originated from the establishment of zoo and garden by Maximilian II in 1569 for using as hunting ground.

EGYPT

이집트는 기원전 약 5000년전 인류 역사상 최초의 국가 형태를 가진 통일 국가를 이루었으며 나일강변을 따라 피라밋을 비롯한 수많은 문화 유적을 만들었다.

Egypt had established the first unified nation in human history about 5000 B.C. and has made numerous cultural heritage along River Nile including Pyramids.

고대 이집트의 최초의 수도는 멤피스였으나 지금의 수도 카이로는 969년부터였다. 카이로는 평균기온이 12.7℃이나 8월의 평균 기온은 27.7℃나 된다.

The first capital of old Egypt was Memphis and the current capital, Cairo had been since 969. The average temperature of Cairo is 12.7℃ but the average temperature in August is 27.7℃.

구루나 마을 전경

Complete view of Gruna Village Cairo

원래 도굴꾼들이 살던 구루나 마을 사람들을 위해 이집트의 명문가 출신 건축가 '하산화티'는 '신구루나' 마을을 설계하고 직접 주민들과 함께 흙벽돌을 찍으며 마을을 만들었다. 이곳에는 사원, 학교, 시장, 집 등의 모든 건축이 이런 모듈화된 흙벽돌에 의해 건축되었고, 우리나라 새마을운동처럼 이집트 사람들에게는 새로운 운동으로 발전하였다.

Gruna was originally a village of rubbers but an architect, 'Hassan Fathy' from excellent Egyptian family had designed and built 'a new Gruna' village for the people in the village by making bricks in person with the residents. In here, all architectures such as temple, school, market, and house were all built in modularized clay bricks and like the new community movement in Korea, this had been developed as a new movement for Egyptians.

REPUBLIC OF SOUTH AFRICA

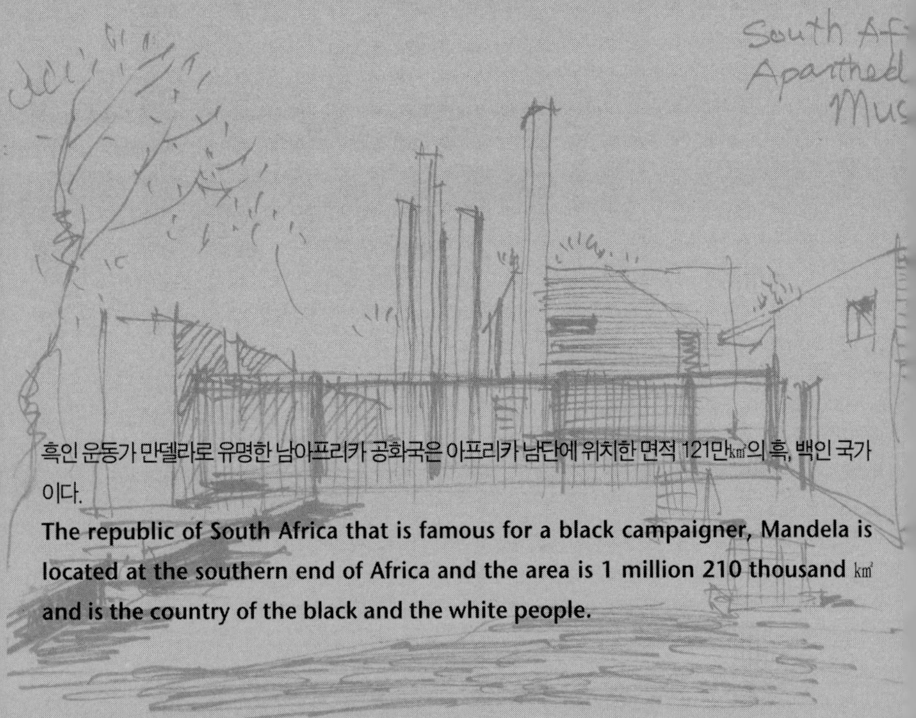

흑인 운동가 만델라로 유명한 남아프리카 공화국은 아프리카 남단에 위치한 면적 121만㎢의 흑, 백인 국가이다.

The republic of South Africa that is famous for a black campaigner, Mandela is located at the southern end of Africa and the area is 1 million 210 thousand ㎢ and is the country of the black and the white people.

요하네스버그

이 나라 최대의 도시이자 아프리카에서 가장 번영한 상공업도시이다. 가장 가까운 항구인 모잠비크의 마푸토에서 철도로 640km, 인도양 연안의 더반항(港)에서 북동쪽으로 650km 떨어진 해발고도 1,900m의 내륙고원에 위치한다. 1886년 금광이 발견되자 황야에 갑자기 건설되었으며, 금광업의 발전과 함께 급격히 팽창하여 1911년에는 유럽인만으로도 케이프타운의 4배나 되는 인구가 되었다.

Johannesburg

The biggest city in South Africa and is the best prospering commercial and industrial city in Africa. It is located at the inland plateau with altitude of 1,900m above sea level 640km away by railroad from the nearest port of Maputo in Mozambique and 650km away in the northeast from Durban port in Indian Ocean. When gold mine was found in 1886, the city was built suddenly and with the development of gold mines, the city had expanded fast and in 1911, population of Johannesburg only with Europeans was 4 times as many as the population of Cape Town.

케이프 타운

배후에 테이블산(1,087m)과 라이온즈 헤드가 솟아 있으며 테이블만(灣)에 면하는 천연의 양항이다. 남아프리카 공화국 의회의 소재지로서 행정부가 있는 프리토리아와 더불어 수도의 지위를 나누어 맡고 있다. 기후는 온대의 지중해성기후로 1월 평균기온 20.3℃, 7월 평균기온 11.6℃이며, 연강수량 526mm로 겨울인 5~8월에 비가 많다. 주민의 약 35%가 백인이며 유럽풍의 대도시 경관을 이룬다. 1652년 4월, 네덜란드 동인도회사가 아시아 무역의 보급기지로서 건설하여 처음에는 네덜란드어로 카프스타드라고 하였으나, 그 후 영국계 식민 활동의 기점이 되었다.

Cape Town

This is a natural good harbor with Table mountain (1,087m) and Lion' s Head in the background and is faced with the Bay of Table. It shares the status of the capital of the republic of South Africa as the location of national assembly with Fretoria that is the location of administration government. Climate is warm Mediterranean weather and average temperature in January is 20.3℃ and 11.6℃ in July and annual rain drops are 526mm and it rains a lot in winter of May~August. About 35% of residents are white and the city has metropolitan European landscapes. It was constructed in April 1652 by the East Indian Company of Netherlands as supply base for trade and was first named as Kaapstad in Netherlands and later became the base for British colonial activities.

케이프타운 해변

Beach in Cape Town Cape Town

남아프리카의 펭귄은 관광객에게 가장 인기 있는 여행 상품이다. 케이프타운 해변에서 펭귄들의 모습은 너무나 귀엽고 낭만적이다.

Penguin in South Africa is the most popular travel product for tourists. Penguin in beaches of Cape Town is so beautiful and romantic.

케이프타운 전경

Cape Town Complete view of Cape Town

우연히 여행을 하면서 만나는 케이프타운의 한 농가 모습.

View of a farm house in Cape Town that we see during travel.

선시티

Sun City Johannesburg

남아프리카를 여행하는 사람들, 특히 건축이나 디자인을 하는 사람들에게 꼭 권하고 싶은 아름다운 곳이 있다. 요하네스버그에서 약 한 시간 정도 거리에 위치해 있는 선시티는 남아공의 라스베가스로 유명한 곳이자 매년 미스남아공 대회를 개최하는 곳이다.
인공의 미와 자연의 미가 완벽하한 조화를 이루는 이 곳은 깊은 산 속에 자리잡은 인공 해변과 웅장한 건축물 등으로 매우 인상적이며 초호화 호텔, 카지노, 각종 오락 시설을 갖춘 완벽한 리조트 시설이 있다.

There is a beautiful place to recommend tourists visiting South Africa, in particular, people in architecture and design business for sure. Sun City located about one hour distance by car from Johannesburg is famous as Las Vegas in the republic of South Africa and every year, beauty contest of Miss South Africa is held here.
Artificial beach seated in deep forest and grandiose buildings are very impressive with the harmony between artificial beauty and natural beauty and it is perfect resort equipped with super luxurious hotel, casino, and various entertainment facilities.

Johannesburg Victoria wharf, Hotel

빅토리아 항구의 모습이다. 바다에 어울린 항구와 건물 사이로 수많은 사람들의 살아가는 모습은 인간적이고 그 자체가 삶이다.

This is a view of Victoria harbor. Features of so many people living between the harbor in harmony with ocean and buildings are so humane and they are life itself.

유럽의 한 항구를 연상케 하는 빅토리아 항구에서의 호텔 모습을 담았다.

A view of a hotel at Victoria harbor that reminds a harbor in Europe.

KENYA

아프리카 대륙 중앙, 적도에 걸쳐져 있고 케냐는 인구 3163만(2003), 면적 58만㎢의 국가로 수도는 나이로비이다.

Kenya located in the middle of African Continent in the equator has population of 31 million 630 thousand people. (In 2003) and the area of 580 thousand ㎢. The capital is Nairobi.

• Nairobi

마사이마라 케냐

Masimara Kenya

마사이마라는 케냐 남서부의 빅토리아호와 그레이트 리프트 밸리 사이에 위치하고 있으며, 근처 탄자니아의 세렝게티 국립공원과 국경선에서 인접해 있으며, 야생 동물의 수가 많기로는 케냐 국내에서 으뜸가는 지구이다.
면적은 1,800㎢로서 제주도와 비슷한 넓이이며 지상에서의 사파리 뿐만 아니라 다른 공원에서는 맛보지 못하는 기구를 타고서 사파리를 즐길 수 있다.

Masimara is located between Lake Victoria and Great Lift Valley in the southwest of Kenya and is faced with Serengeti national park in borderline with Tanzania. This is the place with the highest number of wild animals in Kenya.
The area is 1,800㎢ and is as large as Jeju island and people can enjoy not only safari in land but also in balloon that cannot be found in other parks.

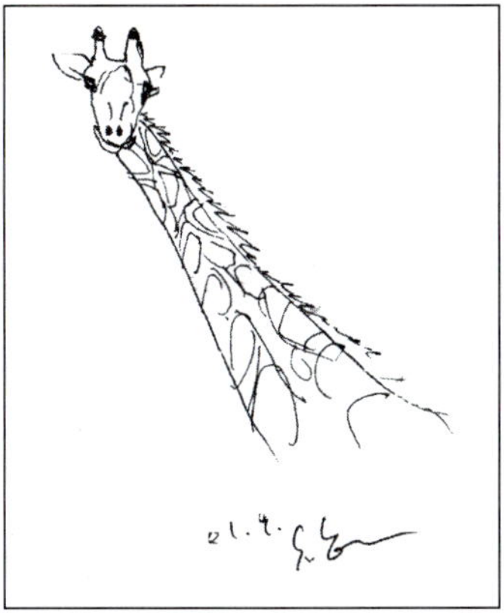

'잠비아' 의 국명은 '잠베지강' 에서 유래되었다. 수도는 루사카이며 공용어는 영국의 식민지의 영향으로 영어를 사용하고 있다. 인구 1000만(2003)으로 면적은 95만㎢ 정도 된다.

The name, 'Zambia' is originated from 'River Zambezi.' . The capital is Lusaka and official language is English due to the influence of British colonialism. Population is 10 million (In 2003) and the area is about 950 thousand ㎢.

ZAMBIA

빅토리아 폭포

Victoria Fall Jimbabwe

1885년 영국의 탐험가 D. 리빙스턴이 발견하였으며, 영국의 빅토리아 여왕의 이름을 따서 '빅토리아 폭포'라 이름을 붙였다. 아프리카의 잠비아와 짐바브웨 국경 경계를 흐르는 세계 최대의 폭포로 해발 90m 고원에서 넓이 1500m, 높이 150m의 낙차로 흘러내린다.

It was discovered by a British explorer, D. Livingston in 1885 and was named as 'Victoria Falls' after British Queen Victoria. This is the largest fall flowing between the border of Zambia and Zimbabwe in Africa and it flows in the width of 1500m at plateau altitude of 90m above sea level and drops in 150m high.

이창근

年譜
Chronological Personal History

Chang Keun, Lee

1943	서울출생
1968	건축 5인전 개최(영천지구 재개발 계획)
1969	인하대학교 건축공학과 졸업
	일양건축연구소 설계실 근무
1973	Lyons Associates 한국지부 건축설계실 근무
1974~현재	(주) 우원디자인 설립, 대표이사
1979~현재	(사) 한국건축가협회 명예이사
1993~1994	(사) 한국실내건축가협회(KOSID) 회장
1995~현재	(사) 한국실내디자인학회(KIID) 명예이사
1995~현재	(사) 한국실내건축가협회(KOSID) 명예회장
1991~2001	대한민국 실내건축대전 심사위원
1992	대한민국 건축대전 초대작가
2000	대한민국 실내건축대전 심사위원장
2005	2005 생활디자인 공모전 심사위원장
2005	대한민국 건축대전 초대작가
1981~1991	명지대학교 건축공학과 출강
1994~2002	경원대학교 실내건축학과 겸임교수
2003~2004	홍익대학교 건축도시대학원 출강
2004~현재	한양대학교 생활과학대학 실내환경디자인전공 겸임교수
2003	그림이 있는 여행 『두 남자의 스케치 · 이야기』 출판전시회(가나아트홀)

■ 수상

1993	서울시 건축상 수상
1994	'94 비룡대상 문화예술부문 수상
1996	'96 KOSID협회상 수상
1998	서울특별시 생활체육 협의회 특별상 수상
2001	서울특별시 성북구청장 표창
	'01 KOSID 협회상 수상
2003	제8회 사단법인 한국실내디자인학회 작품상 수상

스케치와 세계건축여행

Journey of World Architecture & Sketch

스케치와 세계건축여행

Journey of World Architecture & Sketch

2007년 1월 5일 1쇄/2008년6월10일2쇄

지은이 | **이창근** 펴낸이 | **김기현** 펴낸곳 | 도서출판 **시공문화사**

등록 | **1993**년 **3**월 **12**일

120-796 서울시 서대문구 현저동 **200**번지 극동빌딩 **5**층

TEL 02-3147-1212 · 2323 FAX 02-3147-2626

인쇄 **예림인쇄** | 제본 **세진문화(주)** | 용지 **(주) 대림지업사**

www.spacetime.co.kr | **spacetime@korea.com**

5F, 200 Hyeonjeo-dong, Seodaemun-gu, Seoul, Korea

T. 82 0 3147-1212 F. 82 2 3147-2626

15,000 WON